Geology of the Greenock district

The district described in this memoir flanks the estuary of the River Clyde, and extends north to include part of Loch Lomond and south to Largs and Lochwinnoch. The district is covered by Sheet 30W and part of 29E of the geological map of Scotland.

The introductory chapter briefly describes the physical features of the district and outlines its geological development. The oldest rocks in the district, the Dalradian and the Highland Boundary Complex, are described with particular reference to their structure. Chapters then follow dealing with the sediments of Lower and Upper Devonian age, and with the sediments and extrusive rocks of the Carboniferous. Chapters describing the intrusions of various periods and the structural development of the district are followed by an account of the Quaternary deposits, including their history of deposition, and the changes of sea-level which took place during the period. The final chapter is on the economic geology and details the various natural resources present.

BRITISH GEOLOGICAL SURVEY

I B PATERSON, I H S HALL
and D STEPHENSON

Geology of the Greenock district

Memoir for 1:50 000 geological sheet 30W and part of
sheet 29E (Scotland)

CONTRIBUTORS

Geology
I H Forsyth
W G Henderson
J R Mendum
S K Monro

Palaeontology
P J Brand
D K Graham

LONDON: HMSO 1990

iv

© *NERC copyright 1990*

First published 1990

ISBN 0 11 884483 0

Bibliographical reference

PATERSON, I B, HALL, I H S, and STEPHENSON, D.
1990.   Geology of the Greenock district. *Memoir of the
British Geological Survey*, Sheet 30W and part of Sheet 29E
(Scotland).

British Library Cataloguing in Publication Data

A CIP catalogue record of this book is available from the
British Library

Authors

I B Paterson, BSc
I H S Hall, BSc
D Stephenson, BSc, PhD

Contributors

P J Brand, BSc
I H Forsyth, BSc
D K Graham, BA
W G Henderson, BA
J R Mendum, BSc, PhD
S K Monro, BSc, PhD
British Geological Survey, Edinburgh

*Other publications of the Survey dealing with this district and
adjoining districts*

BOOKS

British Regional Geology
The Midland Valley of Scotland, 3rd Edition, 1985
Memoirs
The geology of the Glasgow district, 2nd edition, 1925
The geology of north Ayrshire, 2nd edition, 1930
Economic geology of the Central Coalfield of Scotland,
Area IV, 1920
Memoir for 1:50 000 geological sheet 30E (Glasgow) (in
preparation)

BGS Report
Lithostratigraphy of the late Devonian and early
Carboniferous rocks in the Midland Valley of Scotland.
Vol. 18, No. 3, 1986

MAPS

1:625 000

Geological (North)
Quaternary (North)
Aeromagnetic (Sheet 1)

1:250 000
Clyde (Solid Geology), 1985
Clyde (Sea bed sediments and Quaternary), 1985

1:50 000

Sheet 30W and part of 29E (Greenock   Solid, 1990
Sheet 30W and part of 29E (Greenock)   Drift, 1989

One inch to one mile (1:63 360)
Sheet 29 (Rothesay) Solid and Drift, 1971

Printed in the UK for HMSO
Dd 291142   C10   6/90

CONTENTS

FIGURES

PLATES

TABLES

PREFACE

This memoir provides the first general account of the geology of the area which flanks the lower estuary of the River Clyde and includes the southern part of Loch Lomond. Already known for its outstanding natural beauty, despite its proximity to the major conurbation of Glasgow, the area also possesses a wealth of features of great geological interest. One of the most striking of these is the Highland Boundary Fault-system, which crosses the north-western part of the district. This zone of fractures, which lies along a major tectonic lineament, forms the north-western margin of the sedimentary basin underlying the Midland Valley of Scotland and juxtaposes metamorphic rocks of the Highlands against younger sedimentary and volcanic rocks to the south. Movements on the fault-system profoundly affected the development of sedimentation within the Midland Valley basin, most obviously during Upper Devonian times when uplift to the north maintained a mountainous hinterland capable of sustaining large alluvial fans.

During the early part of the Carboniferous period the influence of the Highland Boundary Fault-system upon sedimentation was more subtle. Application of a newly devised lithostratigraphical classification has revealed how a marine transgression, which had interrupted the deposition of upwards-fining fluvial cycles with pedogenic carbonate upon the coastal floodplain of a major river system, was brought to an end as a consequence of renewed uplift of the Highlands. Subsequently, a new element was introduced into the fill of the early Carboniferous basin by the widespread extrusion of a thick series of mainly basaltic lavas, accompanied by the emplacement of dykes, sills and volcanic necks. Later in the Carboniferous, cyclical sedimentary sequences accumulated. The final form of the landscape clearly displays the effects of repeated glaciation during the Quaternary period.

Beds of coal and limestone of Carboniferous age, and vein-deposits of baryte supported an extractive industry during the 19th and early 20th centuries, but other than hard rock aggregate, the area now has little in the way of exploitable mineral resources. However, a knowledge of the old mine workings is essential when planning for development.

The heavy industries, notably shipbuilding, for which the area around the Clyde estuary was formerly renowned, are in decline, but the area is rapidly acquiring a worldwide reputation as a centre for electronic engineering and other light industries. In addition, there is great scope for the development of the area's potential as a recreational centre. The rock exposures along its coast and streams represent a valuable educational resource. In particular, the opportunities of studying alluvial fan deposits, pedogenic sediments, coastal marine sequences, intrusive and extrusive igneous features and the sedimentary and tectonic structures characteristic of a low-grade metamorphic terrain are almost unparalleled in an area of its size.

Peter J Cook, DSc
Director

23 October 1990

British Geological Survey
Keyworth
Nottingham
NG12 5GG

PREVIOUS SURVEYS

The Greenock district is covered by Sheet 30 of the Geological Map of Scotland and that part of Sheet 29 which lies to the east of the Firth of Clyde. The original survey of these sheets was carried out by A Geikie, J Geikie, R L Jack, E Hull, H M Cadell and J B Hill and the maps were published in Solid and Drift versions at the scale of one inch to one mile in 1878 and 1892 respectively. The ground included in the present district was resurveyed between 1913 and 1954 by E B Bailey, R G Carruthers, C H Dinham, J E Richey, J B Simpson, V A Eyles, A G MacGregor, W Q Kennedy, Dr J G C Anderson, Mr G S Johnstone, Mr G I Lumsden and Mr N R Martin. Solid and Drift editions of Sheet 30 were published in 1958 and 1961 respectively. A combined Solid and Drift edition of Sheet 29, incorporating amendments by Dr M Armstrong, Mr J I Chisholm, Dr G C Clarke and Dr A L Harris, was published in 1971.

What was intended to be a partial revision of the solid rocks of the Greenock district was started in 1976. It became apparent, however, that a complete revision survey of these rocks was necessary and this was completed in 1982. The work was financed in part by the Department of Trade and Industry and, during 1981–82, by the Department of the Environment. The survey was supervised by Messrs E G Poole and J H Hull and was carried out by Mr I H S Hall, Mr W G Henderson, Mr I B Paterson and Dr D Stephenson, with contributions by Dr M Armstrong, Mr I B Cameron, Dr S K Monro and Mr I H Forsyth. No systematic resurvey of the Drift deposits was intended, but amendments to the existing map were made by the surveyors working on the Solid revision and detailed mapping was carried out by Mr M A E Browne in selected areas, notably the Leven valley, in collaboration with Mr J Rose of Birkbeck College, London. Commissioned work on the gravel deposits around Loch Lomond by Mr A M Aitken and Dr M Armstrong in 1984 also resulted in some revision to the Drift lines. Separate Solid and Drift editions of the Greenock Sheet will be published at the scale of 1:50 000.

Neither of the previous surveys of the original one-inch sheets was accompanied by a sheet explanation, although the eastern part of Sheet 30 was described in a special memoir for the Glasgow District. This did not cover any part of the Greenock district. Several papers have been published on various topics concerning the district and adjacent areas since the revision survey and these are referred to in the text. The most notable of these is a revision of the stratigraphical nomenclature of the late Devonian and early Carboniferous rocks, published in BGS Report, Vol. 18 and an assessment of the baryte and copper mineralisation of the Renfrewshire Hills, published as Mineral Reconnaissance Programme Report No. 67.

ACKNOWLEDGEMENTS

The cores and samples from boreholes, drilled by the Geological Survey in 1977–78 to examine the Lower Carboniferous succession and in 1980 to investigate the stratigraphy of the Quaternary sediments, were examined by Dr M Armstrong, Mr M A E Browne, Mr I H S Hall, Mr K I G Lawrie and Mr I B Paterson. The cores from boreholes drilled by civil engineering contractors to investigate foundation conditions for roads, dams, buildings and other works in the area were examined by Mr D N Halley, Mr K I G Lawrie, Mr I B Paterson and Dr D Stephenson.

The Carboniferous macrofauna was revised by Mr P J Brand and Dr R B Wilson, and the Carboniferous macrospores from the Geological Survey boreholes were identified by Dr B Owens. The Quaternary macrofauna obtained from Geological Survey and other boreholes was identified by Mr D K Graham; the microfauna was chiefly identified by Miss D M Gregory and Mr I P Wilkinson. The photographs were mainly taken by Mr T Bain and Mr F I MacTaggart. The memoir was compiled by Mr I B Paterson and edited by Mr J I Chisholm.

ONE

Introduction

The district described in this memoir falls within the Greenock Sheet, which incorporates Sheet 30W of the 1:50 000 geological map of Scotland and the eastern part of 29E. The district flanks the estuary of the River Clyde, extending north to include a small portion of the south-west Highlands and part of Loch Lomond, and south to Largs and Lochwinnoch. It thus contains some of the finest scenery in central Scotland.

The north-western part of the district is crossed by the Highland Boundary Fault-system, a zone of fractures along a major tectonic lineament which forms the north-western margin of the Midland Valley of Scotland and juxtaposes metamorphic rocks of the Highlands against younger sedimentary and volcanic rocks to the south (Figure 1). The differential resistance of these various rocks to erosion is clearly reflected in the nature of the topography. The craggy hills of schist and cleaved greywacke which rise steeply from the ice-scoured fiords of Loch Long and the Gare Loch in the north-west part of the district are typical of the south-west Highlands. South of the River Clyde, the rough uplands and high moors of the Renfrewshire Hills are sculpted from a thick pile of lavas and volcaniclastic sediments of early Carboniferous age. The low ground adjacent to the Clyde coast and elsewhere is, in general, underlain by sandstones, conglomerates and mudstones of Devonian and Carboniferous age. Within the outcrop of these rocks lie by far the greater part of the arable land, most of the centres of habitation and virtually all the industrial development.

During the late Precambrian, when the sediments of which the Dalradian is constituted were laid down, the district was part of the southern margin of the North American continent, which lay on the north side of the Iapetus Ocean. Earth movements, associated with the closure of the ocean by north-westwards subduction, deformed the sediments and converted them into phyllites and cleaved grits. During a later phase of these movements, the assemblage of sediments and ultrabasic igneous rocks known as the Highland Boundary Complex was tectonically emplaced against the continental margin.

Following final closure of the Iapetus Ocean, probably in late Silurian times, the Midland Valley developed as a north-east-trending basin bounded on the north by the Highland Boundary Fault-system. A long series of movements on fractures within this fault-system, particularly during Devonian and Carboniferous times, tended to maintain the relief of the Highlands. This caused coarse detritus to be introduced into the basin, commonly in the form of alluvial fans.

The sedimentary succession within the Midland Valley basin (Figure 2) also shows the effects of tectonic activity, being punctuated by unconformities and nonsequences. The most important of these separates Lower Devonian from Upper Devonian strata, but sedimentation was also interrupted in late Devonian and in mid-Dinantian times. The strata of the Devonian and of the early Carboniferous Inverclyde Group are mainly of fluvial origin, the chief exception being the grey mudstones and dolomitic limestones of the Ballagan Formation which were deposited in coastal sabkhas. The later Carboniferous strata, laid down after the eruption of the lavas of the Clyde Plateau Volcanic Formation, consist mainly of sandstones, mudstones, seatrocks and coals, deposited in part as cyclic sequences in a fluviodeltaic environment. From time to time shelly limestones and mudstones were laid down as a result of marine incursions.

Eruption of the Clyde Plateau Volcanic Formation was accompanied by the emplacement of intrusive igneous bodies of various kinds. Dykes, mainly of doleritic composition, were also intruded during late Carboniferous and Tertiary times.

During the Quaternary, the district was covered on several occasions by ice-sheets which moulded the topography and laid down extensive deposits of till and fluvioglacial sand and gravel.   IBP

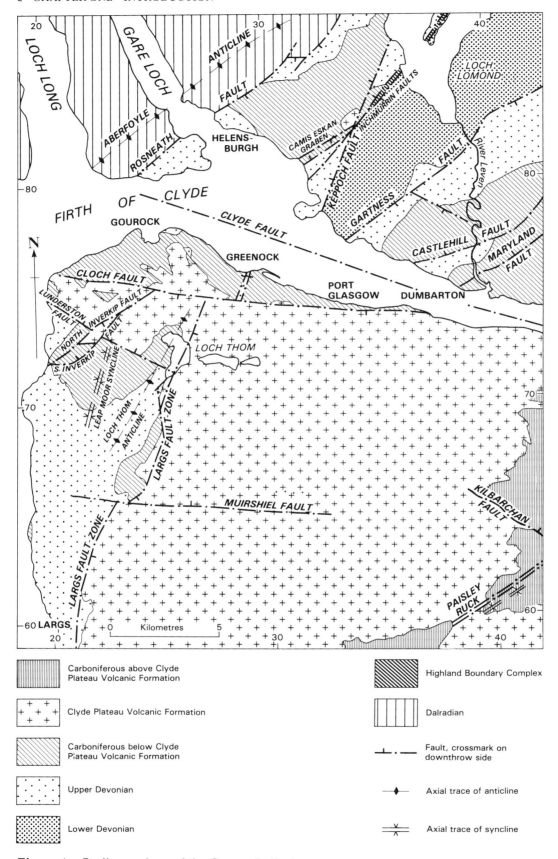

Figure 1 Outline geology of the Greenock district

Figure 2 Generalised sequence of the Devonian and Carboniferous rocks in the Greenock district

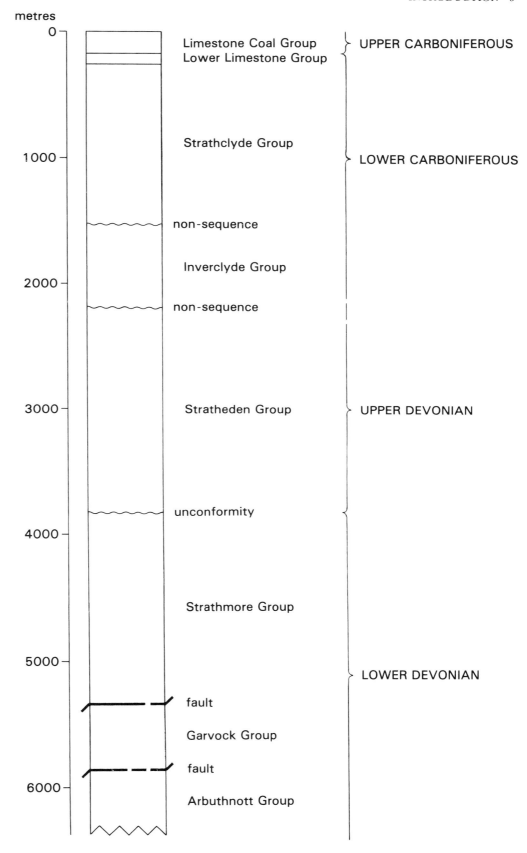

metres

Limestone Coal Group
Lower Limestone Group — UPPER CARBONIFEROUS

Strathclyde Group — LOWER CARBONIFEROUS

non-sequence

Inverclyde Group

non-sequence

Stratheden Group — UPPER DEVONIAN

unconformity

Strathmore Group

LOWER DEVONIAN

fault

Garvock Group

fault

Arbuthnott Group

TWO

Pre-Devonian rocks

DALRADIAN SUPERGROUP

The Dalradian rocks in the district, consisting mainly of phyllites and cleaved greywackes, belong to the Southern Highland Group and are thought to be of late Precambrian age. Three formations have been recognised, the Dunoon Phyllite, the Beinn Bheula Schist and the Bullrock Greywacke, of which the Dunoon Phyllite appears to be the oldest. The Bullrock Greywacke is probably laterally equivalent to part of the Beinn Bheula Schist.

The rocks are mostly of sedimentary origin and were originally composed of muds, poorly sorted sands and fine-grained gravels, considered to have been laid down within a submarine fan complex. The rocks were affected by strong earth movements during the Caledonian Orogeny over a period extending from c.600 Ma to 440 Ma (p.36). They were folded, cleaved and metamorphosed and in this district they are now mainly steeply inclined.

Dunoon Phyllite

The Dunoon Phyllite consists of blue-grey and greenish grey phyllites, black phyllites with lenticular limestone bands, cleaved greywackes and minor fine-grained conglomerates. Most of the greywackes are medium- to coarse-grained sandy units but some are pebbly. The formation is best-exposed along the coast near Cove [221 813] and there are also good exposures in Aldownick Glen [270 850].

On the coast north of Cove, a change from blue-grey phyllites to greenish grey cleaved siltstones marks the contact of the Dunoon Phyllite with the Beinn Bheula Schist. The contact can readily be traced eastwards across the Rosneath peninsula, being offset in places by apparently sinistral movements on north-north-east-trending faults. On the east side of the Gare Loch, the phyllites do not extend as far north as shown on Geological Sheet 29, and the northern limit of the Dunoon Phyllite is now drawn through Croy [260 855]. The southern boundary of the formation is taken at a rapid transition from dark phyllites with limestone lenses to a sequence of coarse, pebbly greywackes, assigned to the Bullrock Greywacke. The contact zone is seen only in Aldownick Glen [270 850].

Greywackes in the Dunoon Phyllite are mainly confined to the southern part of the outcrop and usually occur as separate, graded beds. Load-cast and washout structures are common and the coarser beds in many cases carry siltstone intraclasts in their basal part. Channel bedforms with trough cross-bedding are seen at several places on the coast, for example at a locality [2235 8085] near Barons Point. The pebbly greywackes contain only minor amounts of potash feldspar granules, a feature which distinguishes them from the pink-tinged, more feldspathic pebbly greywackes of the Bullrock Greywacke.

Beinn Bheula Schist

The predominant lithologies in the Beinn Bheula Schist are greenish grey cleaved greywacke and green cleaved siltstone. Generally the greywackes are fine to medium grained in the southern part of the outcrop, near the contact with the Dunoon Phyllite, but farther north they are commonly pebbly, for example in roadside exposures near the north end of Faslane Port. The pebbles are of quartz and feldspar, with a notable proportion of microcline.

Exposures along the coast, at Cove Bay, provide some of the best examples of original sedimentary structures in the Dalradian rocks of the district. The greywacke tends to occur as discrete beds showing Bouma sequences. Repeated grading and ripple-lamination are commonly seen. Calcareous lenses found locally in the greywackes probably formed during diagenesis.

Some beds within the Beinn Bheula Schist are highly chloritic, dark olive-green in colour and have a characteristic massive appearance. These 'green beds' are particularly associated with bodies of metabasite, which occur at Blairmore [195 823] and Knockderry [217 833]. The metabasites are cleaved compact rocks, now composed of actinolite, chlorite, plagioclase and sphene. Some bodies up to 3 m thick, possibly intrusions, clearly show relict igneous textures; others have variable grain size, appear laminated and may be of extrusive origin.

Bullrock Greywacke

The Bullrock Greywacke consists mainly of pink-weathering cleaved feldspathic pebbly greywacke, interbedded with green cleaved siltstone. East of the Gare Loch, black slates with dark grey limestones and blue slates also occur and locally form discrete mappable zones. Formerly the rocks in these zones were ascribed to what is now known as the Highland Boundary Complex, but as they are transitional with typical Bullrock Greywacke lithologies and are enclosed within them, they are now included within the Bullrock Greywacke. Farther south-west in Cowal, black slates and limestones within the Bullrock Greywacke outcrop are placed in a separate lithostratigraphical unit, the Innellan Group.

The Bullrock Greywacke is well exposed on the foreshore from Cove to Kilcreggan, on the hills above Rhu and on both sides of Glen Fruin. The greywackes characteristically occur in amalgamated sequences and do not always show grading. Load-structures and ripple-lamination are seen at some localities. The black slates and limestones are very similar to those within the Dunoon Phyllite. The limestones are recrystallised and occur as lenses about 0.1 m thick.

Plate 1 Coarse- and fine-grained greywackes showing two cleavages, one steep and one subparallel to bedding. Bullrock Greywacke, Kilcreggan Pier [2414 8042] (D 2602).

Conditions of deposition

The sedimentary character of most of the Dalradian rocks is consistent with deposition in submarine fans by turbidity currents and mass flow. In terms of the models proposed for submarine fans by Middleton and Hampton (1973) and Mutti and Ricci-Lucchi (1972), it is inferred that the Bullrock Greywacke was laid down in an inner- or mid-fan environment. The portion of the Beinn Bheula Schist which is present in the Greenock district may have been deposited on the outermost parts of a fan. The Dunoon Phyllite appears to have been laid down in a more distal situation, perhaps in an interfan environment.

Structure

The Dalradian rocks lie within the Tay Nappe, a major recumbent F1 structure which dominates the Southern Highlands. The limbs of the nappe are generally flat-lying and the structure faces to the south-east. Its hinge-zone, known as the Aberfoyle Anticline, lies within the Greenock district. A later phase of deformation has 'down-bent' the hinge-zone relative to the horizontal limbs, so that the Aberfoyle Anticline is now synformal and lies within a 'steep belt' in which the bedding and cleavages generally dip steeply towards the south-east. This steep belt is some 10 km wide immediately north-west of the Highland Border and includes all the Dalradian rocks of the Greenock district.

Because of the present attitude of the Aberfoyle Anticline, strata on its north-west limb are generally inverted whereas beds on the south-east limb are mainly the right way up. Cleavages are generally downward-facing, that is, the beds become younger down dip on the cleavage planes. An overall change in younging direction takes place close to the south-east margin of the Dunoon Phyllite, inverted beds predominating to the north-west and right-way-up strata being more usual to the south-east. It is considered that the south-east margin of the Dunoon Phyllite is formed by an early slide which disrupts the hinge-zone of the Aberfoyle Anticline.

Minor folds, cleavages and related lineations are associated with three distinct regional deformation phases (D1, D2 and D4). Tight to isoclinal minor folds (F1), typically with axes plunging at low angles to the south-west or the south-south-west, are seen in quarries and coastal exposures. Axial planar to the folds is a pervasive primary cleavage (S1) which in the greywackes is developed both as a discrete pressure solution cleavage (S1p) and as a subparallel grain alignment (S1g), principally of phyllosilicate minerals. In pelitic lithologies, the only S1 cleavage to be detected is a slaty cleavage, which is present both on fold limbs and in hinge zones. At lower structural levels, in the north-western part of the Dalradian outcrop, the intensity of the D1 deformation increases.

A later deformational event has superimposed a second series of minor folds and variably developed cleavages upon the D1 structures. Both the bedding and S1 are deformed by F2 folds which typically verge towards the north-west. The axes of these folds plunge at moderate angles towards the south-west. In the area to the SSE of the Dunoon Phyllite outcrop, the secondary cleavage (S2) occurs as a weakly developed fracture cleavage or as a spaced pressure solution cleavage in the greywackes and as a crenulation cleavage in the phyllites. Within and immediately NNW of the Dunoon Phyllite outcrop, S1 and S2 may combine to form a composite fabric which is represented on the map as a single cleavage. Still farther north-west, at deeper structural levels, S2 is the dominant fabric. Typically it trends north-east to NNE and dips moderately south-eastwards. Where distinguished as a separate structure, it occurs as a discrete, spaced, pressure solution cleavage, locally with associated close to tight minor folds. In pelitic horizons, S2 is a slaty or strong crenulation cleavage which modifies or overprints the S1 fabric.

As is seen in much of the south-west Highlands, a persistent stretching lineation which plunges to the south-east or ESE is manifested by elongated grains and fibrous mineral growths on S1 and S2 surfaces (see also Gunn et al., 1897; Roberts and Treagus, 1977). This lineation makes a high angle with the fold axes and cleavage/bedding intersections.

All the Dalradian rocks in the district have been affected by a still later deformational event, which, by correlation with the structural sequence in the adjoining Ben Lomond district (Sheet 38W), is termed D4. This D4 compression largely postdates the formation of the monoformal 'downbend' (p.5), itself attributed to the first stages of block uplift of the Dalradian terrain. Within the district, small-scale crenulations and related cleavages of D4 are only weakly developed at a few localities. A weak, subvertical, south-south-east-trending crenulation cleavage on the foreshore at Loch Long and a weak crenulation cleavage in the finer-grained greywackes at Kilcreggan may well be a product of D4. They may, however, relate to movements on fractures in the Highland Boundary Fault-system, which continued to be active during Devonian and Carboniferous times.

HIGHLAND BOUNDARY COMPLEX

A narrow belt of serpentinitic rocks extends south-westwards from Inchmurrin in Loch Lomond to Ben Bowie [340 828]. These rocks belong to the Highland Boundary Complex, an exotic assemblage of igneous and sedimentary rocks in part of ocean-floor origin that can be traced from Stonehaven on the North Sea coast to the Isle of Arran and into Northern Ireland. Over most of this distance the Complex occurs along the south-eastern margin of the Southern Highland Group (Upper Dalradian) but at Loch Lomond, Upper Devonian rocks are faulted down between it and the Dalradian. Fossil evidence from sedimentary rocks of the Complex at Aberfoyle and elsewhere (see Curry et al., 1984) indicates an Ordovocian age.

In this district, the Highland Boundary Complex comprises serpentinite, mostly altered to carbonate and silica minerals, and fragmental rocks made up mainly of altered serpentinites. On Inchmurrin and Ben Bowie, psammitic clasts within an altered serpentinitic matrix may represent conglomeratic sediments like those found at Dounans near Aberfoyle. Steeply dipping faults, some perpendicular to the trend of the Highland Boundary Complex outcrops, others intersecting the trend at low angles, bring different lithologies against one another. Evidently the Highland Boundary Complex was finally emplaced into its present position as the result of brittle fault-movements at the Highland Boundary.   WGH, JRM

THREE

Lower Devonian

Lower Devonian rocks occur in a north-east-trending zone, about 5 km wide, which extends from the River Clyde to the shores of Loch Lomond (Figure 1). The outcrop is bounded to the north-west by the South Inchmurrin Fault, an element in the Highland Boundary Fault-system (Figures 1 and 8) and to the south-east, along most of its length, by the Gartness Fault. The strata generally dip to the south-east and form part of the north-west limb of the Strathmore Syncline, a structure that extends along the entire Midland Valley. In the Greenock district the axial zone and the south-east limb of the syncline are buried below younger sediments. The sequence is a continuation of that described for the north-west limb of the Strathmore Syncline in the Stirling district by Forsyth (*in* Francis et al., 1970, p.66) and Armstrong and Paterson (1970). The strata are not well exposed and are considerably affected by faulting so that only minimum thicknesses can be estimated for the various units (Table 1). These minimum thicknesses, however, provide little support for the suggestion (Qureshi, 1970; Morton, 1979) that the various units are markedly thinner than their equivalents in areas to the north-east.

The unconformable base of the Lower Devonian is not seen in the district but to the north-east, at Aberfoyle (Curry et al., 1984) and Comrie, and to the south-west on Arran, the oldest strata present almost certainly belong to the Arbuthnott Group. This is probably true also of the present district, the lower part of the succession recognised in Kincardineshire having been overlapped by younger Lower Devonian strata.

Table 1 Sequence of Lower Devonian strata

Lithostratigraphical units	Thickness (metres)	Depositional environment
Strathmore Group:		
Teith Formation	1000	FP/MR
Cromlix Formation	500 +	PL or DAF
Garvock Group:		
sandstones	500 +	FP/BR
Arbuthnott Group:		
Inchmurrin Conglomerate	500 +	AF

Depositional environments (partly after Morton, 1979; Mykura, 1983; and Armstrong et al., 1985)

AF	Alluvial fan
DAF	Distal alluvial fan
FP/BR	Floodplain, dominantly braided rivers
FP/MR	Floodplain, dominantly meandering rivers
PL	Playa lakes

ARBUTHNOTT GROUP

The oldest known Lower Devonian strata in the district are conglomerates and pebbly sandstones, the Inchmurrin Conglomerate, which are correlated with the conglomerates seen at Balmaha and Callander. They have been placed in the Arbuthnott Group. Along most of its length, the outcrop of the Inchmurrin Conglomerate is bounded on the north-west by rocks of the Highland Boundary Complex, but in places [as at 350 832] pebbly sandstone occurs in fault-slices interposed between the conglomerate and the older rocks.

The Inchmurrin Conglomerate is well sorted and consists of well-rounded pebbles, cobbles and boulders up to 0.2 m in diameter set in a coarse sandstone matrix. There are also some beds of pebbly sandstone at higher levels. Consistently throughout the outcrop, the clasts in the lower beds are dominantly quartzite, similar to Dalradian quartzites on Jura (Friend et al., 1963) and Schiehallion. In higher beds these are mixed with clasts of acid and intermediate volcanic rocks. Near the Highland Boundary Fault-system, the quartzite clasts have been deformed by brittle fracture, probably in Middle Devonian times (Ramsay, 1964), and in places show the effects of pressure solution where one clast impinges upon another. The fractured segments have subsequently been cemented together and the rock as a whole is very hard and resistant to weathering. Because of their size and well-rounded state, it has been suggested that the quartzite clasts are polycyclic, having been laid down and eroded more than once during transport to their present site (Bluck, 1978).

GARVOCK GROUP

Sediments of the Garvock Group are very poorly exposed in the district. In the Callander area, where they are estimated to be 2000 m thick, they consist of conglomerate and sandstone. The conglomerates which make up more than half the thickness of the group near Callander die out towards the present district but farther to the south-west, in Arran and Kintyre, conglomerate again becomes the dominant lithology.

In the present district, Garvock Group sediments are exposed only along the coast south of Ardmore. The sequence consists mainly of purple-red and grey-brown sandstones with a few beds containing small quartz and Highland pebbles. Subordinate siltstones also occur and in some cases are very micaceous.

STRATHMORE GROUP

Two formations of the Strathmore Group, the Cromlix Formation and the overlying Teith Formation, are represented in the present district. The Cromlix Formation consists predominantly of poorly bedded, poorly sorted, silty mudstone. Siltstones and sandstones occur west of Loch Lomond and it appears that the mudstones pass into sandstones to the south-west as well as to the north-east. The formation was not identified by Francis et al. (1970) on the north-west limb of the Strathmore Syncline in the Callander area. However, Armstrong and Paterson (1970) considered that it was represented by the 'lower purple sandstone facies', the oldest of three divisions recognised in the Teith Formation in this area by Francis et al.

The Teith Formation occupies by far the greater part of the Lower Devonian outcrop in the district. It consists mainly of sandstones with subordinate mudstones and siltstones in upward-fining fluvial cycles. The sandstones, which are purple or brown-grey in the lower part of the formation and pale grey-brown or yellow-grey at higher levels, commonly have mudstone clasts at the base. They are considered to have been laid down within river channels. The mudstones and siltstones, which are dark brown, purple or grey, in a few cases contain plant debris. Bands of carbonate concretions also occur in the mudstones, which are considered to represent overbank sediment laid down on floodplains. IHSH

Species of the genera *Drepanophycus* and *Sawdonia* have been recorded in the present district from Auchensail Quarry [342 798] (Scott et al., 1976) and *Psilophyton* has been found at the same horizon farther east. This suggests a lower Emsian age for these strata. IHSH, PJB

East of the Leven valley the highest beds of the formation show considerable reddening, suggesting proximity to the unconformity with the Upper Devonian. Reddening is also seen west of the Leven below the Upper Devonian conglomerates on Overtoun Muir [363 798].

CONDITIONS OF DEPOSITION

It is considered that deposition of the Lower Devonian sequence took place in an elongate, north-east-trending basin, centred on the Midland Valley and bounded on the north-west by the Highlands (Bluck, 1978; Morton, 1979; Mykura, 1983; Armstrong et al., 1985). Material derived from the Highlands was laid down at the basin margin in alluvial fans and partly dispersed along the basin by axial drainage from the north-east. The quantity and coarseness of the fan material, represented by the conglomerates in the Arbuthnott and Garvock groups, suggests that deposition of these formations took place in the vicinity of scarps controlled by major growth faults. Erosion in Lower Devonian times, following uplift to the north on faults now concealed beneath later Lower Devonian strata, may partly explain the absence of the Dunnottar and Crawton groups and part of the Arbuthnott Group.

While the finer-grained parts of the Garvock and Strathmore groups were being laid down, it is clear that the basin margin had migrated some distance to the north-west, to a position within the south-east Highlands. Within the limits of the present outcrop, deposition was mainly by braided river systems in the regime of the axial drainage. During deposition of the Cromlix Formation, which has been interpreted as a playa lake deposit (Mykura, 1983) or as distal alluvial fan sediment (Armstrong et al., 1985), the axial component of the drainage would appear to have greatly diminished, possibly as a result of more arid climatic conditions. Axial drainage was re-established during deposition of the Teith Formation, as a result either of renewed uplift of the Highland source areas or of increased rainfall; the environment was predominantly one of meandering streams on a broad floodplain. IHSH

FOUR

Upper Devonian

STRATHEDEN GROUP

Strata consisting mainly of red-brown, commonly pebbly, cross-bedded sandstones and subordinate conglomerates formerly referred to the Upper Old Red Sandstone are now assigned to the Stratheden Group (Paterson and Hall, 1986). No fossils have been found in these rocks in the Greenock district but assemblages of fossil fish found in similar strata elsewhere in the Midland Valley indicate an Upper Devonian (Famennian) age. The base of the Stratheden Group is not seen within the district but in adjacent areas there is a marked unconformity on rocks of Lower Devonian age or older. The top of the group is taken at the base of the Kinnesswood Formation, a sequence of cornstone-bearing sandstones and silty mudstones previously placed in the Upper Old Red Sandstone. Locally, this junction is sharp and may mark a non-sequence. The strata of the Stratheden Group show considerable variation within the district, as shown in Table 2.   IBP, IHSH

Wemyss Bay Sandstone Formation (WEM)

The oldest strata south of the Clyde estuary and west of the Largs Fault-zone are assigned to the Wemyss Bay Sandstone Formation. They consist of red-brown, fine-grained, pebble-free sandstone exposed in a faulted anticlinal structure at Wemyss Bay [189 699], and in places along the coast southwards beyond Skelmorlie. The base of the formation is not seen. The sandstones show large-scale, low-angle, bipolar cross-bedding, and may be of aeolian origin. The dominant cross-bedding indicates transport from the south-south-east (Bluck, 1978). At Wemyss Bay the formation is overlain, with an erosive contact, by the coarse-grained Skelmorlie Conglomerate (Plate 2).

Skelmorlie Conglomerate (SKM)

This formation is composed of well-rounded to subangular pebbles and boulders up to 0.15 m in diameter of quartzite, quartz, low-grade schist, sandstone and lava. Low-angle cross-bedding is present in places and the clasts commonly show an imbricate arrangement. In the upper part of the formation, the grain size diminishes, there are beds of coarse pebbly sandstone and the conglomerate passes gradationally into the overlying Kelly Burn Sandstone Formation.

The Skelmorlie Conglomerate is considered to have been laid down as an alluvial fan deposit by braided rivers. The clast composition (Bluck, 1980, fig. 5) indicates that, in all probability, the source rocks consisted of conglomerates and lavas or lava-conglomerates of Lower Devonian age, in addition to Dalradian metasediments.

The full thickness of the conglomerate (20 m) is exposed on the coast at Wemyss Bay [190 696]. The formation is also seen along the shore at Skelmorlie and in the cliffed back feature of the low coastal platform.

Kelly Burn Sandstone Formation (KBS)

The great bulk of the Stratheden Group south of the River Clyde is assigned to the Kelly Burn Sandstone Formation. The formation is composed of medium- to coarse-grained, trough-cross-stratified sandstone, commonly pebbly, with subordinate amounts of conglomerate (Plate 9). The pebbles within the sandstones and conglomerates are usually angular to subangular and consist mainly of quartz and low grade metamorphic rocks with lesser amounts of quartzite and lava. Locally, as on the shore at Gourock [236 774] where the uppermost part of the sequence is conglomeratic, and in Noddsdale [2742 6566], north of the reservoir at Outerwards, the proportion of lava clasts is greater.

Channel forms are commonly preserved in the sandstones, good examples being seen along the coast north of Largs. Clasts of red-brown silty mudstone are common at the bases of such channel fills, as well as the more usual quartz and schist pebbles, but only rarely have intact beds of mudstone survived. In a few places, north of Largs, the bedding is disturbed by water-expulsion structures. The Kelly Burn Sandstone Formation as a whole is considered to have been laid down by a braided river system, which, as indicated by the cross-bedding directions, flowed generally towards the north-east. The source-terrain of the sandstones apparently was composed largely of Dalradian metasediments which,

Table 2   Formations in the Stratheden Group

North of the Clyde estuary	South of the Clyde estuary	
Helensburgh–Dumbarton area	West of Largs Fault-zone	East of Largs Fault-zone
Kinnesswood Formation (base of Inverclyde Group)		
Stockiemuir Sandstone Formation	?non-sequence	Fairlie Sandstone Formation
	Kelly Burn Sandstone Formation	Seamill Sandstone Formation
	Skelmorlie Conglomerate	
Rosneath Conglomerate	Wemyss Bay Sandstone Formation	

Plate 2 Erosive contact of Skelmorlie Conglomerate on Wemyss Bay Sandstone Formation. Wemyss Bay [188 699] (D 1557).

however, were still partly covered by Lower Devonian rocks or by strata deposited earlier in the Stratheden Group period. The formation is well exposed on the foreshore north of Largs, at Skelmorlie and between Wemyss Bay and Lunderston Bay. There are good sections in many streams, the best being along the Kelly Burn [193 684 to 223 684] and the Skelmorlie Water [195 657 to 213 661].   IBP

Seamill Sandstone Formation (SEA)

East of the Largs Fault-zone, the sandstones of this formation crop out only in small areas [210 600] on the lower hillslopes east and north-east of Largs. They are red-brown or brown in colour, generally fine grained and contain scattered quartz pebbles and a few thin beds of quartz-pebble conglomerate. The sandstones are usually flat bedded but in some cases show large-scale, low-angle cross-bedding. The formation is best exposed south of the present district, along the shore between Seamill and Fairlie, and in the glens of

Fairlie and Kelburn, where it passes gradationally into the succeeding Fairlie Sandstone Formation.

Fairlie Sandstone Formation (FAS)

In the present district, the Fairlie Sandstone Formation occurs only in a small area [218 607] north of Largs, where it consists of white or cream-coloured fine-grained sandstone. The formation is well developed south of Largs in the glens of Fairlie and Kelburn. Here, beds of cross-laminated sandstone with scattered quartz pebbles and green siltstone clasts alternate with beds in which the sandstones display a close flat lamination similar to that which characterises the Knox Pulpit Formation of central Fife (Chisholm and Dean, 1974). There are also a few thin beds of conglomerate in which the pebbles consist mainly of vein quartz with a few quartzites. In the Largs Borehole, white cross-bedded pebbly sandstones lying beneath cornstone-bearing strata of the

Kinnesswood Formation, may belong to the Fairlie Sandstone Formation.   DS

Rosneath Conglomerate (RON)

In the Helensburgh–Dumbarton area, the Rosneath Conglomerate is well exposed along the foreshore between Rosneath Point [275 805] and Kilcreggan. In its lower part, the formation is composed of poorly sorted conglomerate and pebbly sandstone in planar units mainly 0.3 to 0.6 m thick (Plate 3). The angular to subrounded clasts consist predominantly of quartz and low-grade schist with lesser amounts of quartzite. Near Rosneath Point there is a marked unconformity, above which the conglomerate is much coarser grained, poorly bedded, and contains numerous boulders of granite, 0.3 to 0.6 m in diameter, in addition to the usual content of quartz, schist and quartzite. Isolated boulders of granite occur throughout this part of the sequence. In the uppermost part of the exposed sequence, the clasts tend to be smaller, the units thicker and the proportion of pebbly sandstone greater. Cross-bedding measurements carried out in the present survey are in accord with the observation by Bluck (1980) that the conglomerates were laid down by palaeocurrents flowing towards the north-east. However, at least some of the thinly bedded conglomerates in the lower part of sequence at Rosneath Point were deposited by a drainage directed towards the south-east.

On earlier maps of the Greenock district, conglomerates at Cardross [335 774] and Overtoun Muir [370 800] and some of those on Ardmore peninsula [315 785] were placed in the Lower Old Red Sandstone. At all these localities, the conglomerates are poorly sorted and their clast content consists mainly of angular and subangular fragments of vein quartz and low-grade schist with subordinate amounts of well-rounded quartzite boulders. In these respects they more closely resemble the Rosneath Conglomerate than any Lower Devonian rocks in the neighbourhood. The quartzite clasts are considered to have been derived from Lower Devonian conglomerates and the fact that a number of them have been split and subsequently abraded is evidence that they have been recycled. Accordingly, the conglomerates in the three areas are now assigned to the Rosneath Conglomerate. Conglomerates to the north-east of Helensburgh contain a higher proportion of angular and slabby clasts mainly of metasedimentary rocks, apparently of local origin. Pebbly sandstones are largely confined to the higher beds.

Stockiemuir Sandstone Formation (SCK)

The lower part of the Stockiemuir Sandstone Formation consists of red-brown fine-grained cross-bedded quartzose relatively mature sandstones with a few pebbly beds and rare

Plate 3 Water-laid conglomerate and cross-bedded sandstone. Rosneath Conglomerate, Rosneath Point [274 806] (D 3596).

mudstone layers. In the upper part of the sequence, these water-laid sediments are interbedded with cross-bedded sandstones (Plate 4) considered to be of aeolian origin (Hall and Chisholm, 1987). These sandstones lack mudstone or siltstone intraclasts, extrabasinal pebbles, parting lineations and micaceous films on bedding planes, features typical of the water-laid deposits, but show a fine 'pin-stripe' lamination of the type considered by Fryberger and Schenk (1988) to be characteristic of aeolian deposition. The trace fossil *Planolites* has been reported at two localities [388 760 and 398 796] (Aspen, 1974) and *Skolithos*-like burrows have been recorded from the same areas (Hall and Chisholm, 1987). The burrows are believed to have formed in wet interdune areas.

The relationship of the Stockiemuir Sandstone Formation with the Rosneath Conglomerate is not seen because of faulting but it is assumed that the sandstones overlie the conglomerates and partly replace them laterally.　　IHSH

CONDITIONS OF DEPOSITION

As a result of regional uplift associated with Middle Devonian earth movements, a major east-north-east-trending intermontane basin was eroded across central Scotland (Figure 3). In the area adjacent to the Firth of Clyde, a number of tributary valleys appear to have entered the basin from the west and north. In Devonian times, regional subsidence allowed sediment to accumulate in the channels of mainly braided river systems which flowed along the axes of the main basin and its tributaries, as is indicated by the evidence of clast size distribution and palaeocurrent directions. During the later part of the Upper Devonian, aeolian sediments were laid down in places, particularly along the axis of the main basin.

The relatively mature Seamill Sandstone Formation, which was laid down by a generally northwards-directed drainage, and the water-laid parts of the Stockiemuir Sand-

Plate 4　Aeolian cross-bedding in Stockiemuir Sandstone Formation, exposed in cliff at back of Main Postglacial Shoreline. Havock Hole, near Dumbarton [380 756] (D 3605).

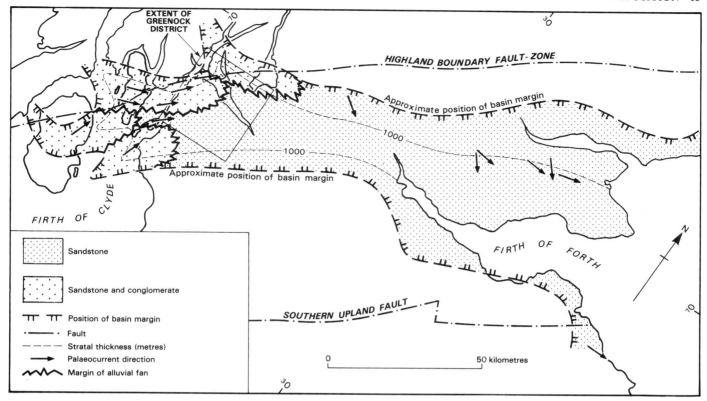

Figure 3   Reconstruction of Stratheden Group palaeogeography in central Scotland. Data partly from Bluck (1978; 1980), Chisholm and Dean (1974) and Read and Johnson (1967).

stone, probably represent the deposit of the axial drainage of the main basin. By contrast, the coarse lithic arenites of the Kelly Burn Sandstone Formation, the Skelmorlie Conglomerate, and the bulk of the Rosneath Conglomerate, are considered to have been laid down in the more distal parts of a major alluvial fan which entered the Greenock district from the west, by way of Bute (Figure 3). The angular and subangular vein quartz and schist clasts which form the chief component of these deposits indicate that the source terrain was composed principally of Dalradian metasediments. However, the generally well-rounded boulders of quartzite, lava and granite are considered to have had their source in conglomerates of Lower Devonian age, as there are no other rocks of suitable composition in the area upstream of the fan. South-eastward palaeocurrents inferred for the Ardmore conglomerates and some beds in the lower part of the Rosneath sequence indicate deposition in another alluvial fan, which entered the main basin from the north.

The persistence of conglomerate in the Stratheden Group in the north-western part of the Greenock district suggests that the relief of the Highland source areas which supplied the alluvial fans was maintained by contemporaneous movements on faults, probably fractures within or parallel to the Highland Boundary Fault-system. It has been suggested that these fault-movements involved large-scale lateral displacements (Bluck, 1978; 1980) but there is no evidence which unequivocally supports this view.

During deposition of the Stratheden Group, the rainfall was probably moderate and somewhat seasonal but the presence of presumed aeolian sediments in the Stockiemuir and Fairlie sandstones suggests that the rainfall and the relief may have diminished through the period. These formations were probably laid down throughout the entire Greenock district but prior to deposition of the succeeding Inverclyde Group were removed by erosion from the area west of a line passing through Largs and Greenock, as a result of uplift on the Largs Fault-zone.   IBP, IHSH

FIVE

Carboniferous

The stratigraphical classification of the lower Carboniferous rocks (below the Hurlet Limestone) in the Midland Valley of Scotland has recently been revised (Paterson and Hall, 1986). Rocks formerly included in the Calciferous Sandstone Measures, together with an underlying sequence of sandstones and siltstones with pedogenic carbonate at the top of the Upper Old Red Sandstone, are now assigned to two new units, the Inverclyde Group and the Strathclyde Group (Table 3). For strata above the Hurlet Limestone the existing classification is retained, except in the Lochwinnoch area, where a local succession is employed (Table 3).

INVERCLYDE GROUP

In much of the district, the Inverclyde Group comprises the Kinnesswood, Ballagan and Clyde Sandstone formations (Paterson and Hall, 1986). The mainly arenaceous Kinnesswood Formation, at the base of the group, was formerly regarded as part of the Upper Old Red Sandstone. However, it is readily distinguished from the strata of the Stratheden Group by the presence of nodules and beds of concretionary carbonate ('cornstone') of pedogenic origin, a feature which it shares with parts of the Clyde Sandstone Formation. The Ballagan Formation consists of a distinctive sequence of grey silty mudstones and numerous thin beds of mainly dolomitic limestone ('cementstone'). The base of the Kinnesswood Formation is generally sharp and may mark a nonsequence. The upper and lower boundaries of the

Ballagan Formation are transitional and almost certainly diachronous.   IBP, IHSH

The Inverclyde Group contains few fossils of zonal value. Disarticulated scales of the fossil fish *Bothriolepis* and *Holoptychius*, found in the basal conglomerate of the Kinnesswood Formation at Fairy Knowe Quarry [369 789], near Dumbarton (Aspen, 1974), are of Famennian age but they may have been reworked from older deposits. The grey mudstones of the Ballagan Formation commonly contain a moderately abundant fauna of ostracods, fish scraps and *Modiolus,* and *Lingula* was found at a single locality [419 785] near Renton. Miospore assemblages recovered from argillaceous beds in the Ballagan and Clyde Sandstone formations, penetrated by BGS boreholes at Barnhill and Loch Humphrey, sited a little to the east of the present district, are considered by Dr B Owens to indicate an upper Tournaisian age. Hence it is probable that at least part of the Kinnesswood Formation is of Lower Carboniferous age, although its basal part may be Upper Devonian.   PJB

Kinnesswood Formation (KNW)

The strata of the Kinnesswood Formation consist of beds of red and white sandstone and red-brown, green-spotted silty mudstone arranged in upward-fining fluvial cycles. In such cycles, the coarse-grained lower part is considered to have been deposited in the channels of a river system; the fine-grained upper part represents 'overbank' sediment, laid down upon the associated floodplains.

Table 3 Stratigraphical framework of the Carboniferous rocks

Age	General sequence			Lochwinnoch area
Namurian	Limestone Coal Group			Dalry Sandstone Formation
				Kilbirnie Mudstone Formation
Dinantian	Lower Limestone Group			Lugton Limestone Formation
	Strathclyde Group	Lawmuir Formation		Kirkwood Formation
		Kirkwood Formation		
		Clyde Plateau Volcanic Formation		Clyde Plateau Volcanic Formation
	Inverclyde Group	Clyde Sandstone Formation	local non-sequence	?
		Ballagan Formation		
		Kinnesswood Formation		

The sandstone beds are mainly fine to medium grained, generally cross bedded and have sharp erosive bases. In the lower part of many sandstones there are angular clasts of mudstone and limestone derived by erosion from the overbank deposits, as well as quartz and quartzite pebbles of extrabasinal origin. Many of the sandstones comprise several channel-fill units, each with an erosive base. Where the cycles contain a single sandstone unit, this is usually from 2 to 3 m thick, but multistorey sandstones may reach a thickness of 10 to 15 m.

The red-brown silty mudstone overbank deposits range in thickness from a few centimetres to several metres, and may contain thin beds of fine-grained sandstone. Sand-filled desiccation cracks are common, in some cases forming polygons, and may penetrate to depths of more than a metre.

Carbonate, (mixtures largely of calcite and dolomite, see Table 4 and Figure 4), occurs widely as irregular nodules (glaebules), which were secreted within developing soil profiles (Plate 5). The glaebules may be scattered sporadically throughout both the coarse- and fine-grained components of the fluvial cycles but are more usually concentrated in a zone near the top of the overbank sediment. Where the overbank deposit is thicker than average, several calcareous horizons may occur, suggesting local persistence of the floodplain conditions while a number of cycles developed elsewhere. In more mature soil profiles, the glaebules tend to be elongated in a vertical sense and to lie along the planes of the polygonal jointing. Particularly good examples of this are seen in the coast sections at Auchengarth and at the Inverkip War Memorial. At the latter site [201 719], a 2 m-thick bed of

Table 4 Determinations of various metals in the acid-soluble fractions of carbonate-rich sediments in the Inverclyde and Stratheden groups.

Sample No.	Ca%	Mg%	Ca/Mg	Fe (ppm)	Mn (ppm)	Sr (ppm)	Ba (ppm)	Depth (m)	LISC
Kipperoch Borehole									
FX 211	19.1	10.6	1.80	5000	3200	320	90	10.53	BGN
FX 212	17.5	10.4	1.68	6100	2200	1150	140	13.47	BGN
FX 213	13.6	6.4	2.13	6000	1800	170	10	27.28	BGN
FX 214	14.0	8.1	1.73	7400	1800	150	1800	27.55	BGN
FX 215	6.3	1.1	5.73	10800	170	<50	25	49.04	BGN
FX 216	15.8	9.4	1.68	5100	1370	150	20	50.90	BGN
FX 217	12.5	7.7	1.62	5500	960	450	25	51.33	KNW
FX 218	13.7	10.3	1.33	3100	1310	300	50	74.53	KNW
FX 219	17.2	8.7	1.98	5000	1100	95	20	77.20	KNW
FX 220	18.4	10.4	1.77	900	1500	110	10	95.87	KNW
FX 221	15.9	8.4	1.89	760	860	70	<10	121.62	KNW
FX 222	12.8	6.9	1.86	550	670	60	20	121.81	KNW
FX 223	13.4	6.8	1.97	540	640	60	30	122.01	KNW
FX 224	11.4	6.5	1.75	570	600	60	25	122.20	KNW
FX 225	13.9	7.3	1.90	1140	1360	60	25	192.80	KNW
FX 226	16.3	4.3	3.79	610	1700	70	45	217.96	KNW
FX 227	3.5	0.1	35.00	460	340	30	15	223.58	SCK
FX 228	3.8	0.1	38.00	540	420	30	45	231.36	SCK
FX 229	7.6	0.3	26.00	490	720	50	30	234.10	SCK
FX 230	2.6	0.1	2.14	730	430	30	250	279.26	SCK
Everton Borehole									
ZH 1267	17.1	8.0	2.14	8100	2000	310	60	22.35	BGN
ZH 1268	23.1	9.9	2.33	2400	2200	520	100	36.70	BGN
ZH 1269	17.4	8.1	2.15	6500	1500	390	1600	39.35	BGN
ZH 1270	14.9	8.2	1.82	1200	1200	260	45	43.60	BGN
ZH 1271	31.8	0.4	75.50	840	260	260	15	108.70	KNW
ZH 1272	26.5	1.2	22.08	1150	620	250	10	113.85	KNW
ZH 1273	29.6	0.4	74.00	760	760	350	15	120.80	KNW
Knocknairshill Borehole									
ZH 1274	29.9	0.3	99.67	1300	1000	410	180	138.90	CYD
ZH 1275	21.3	0.4	53.25	3100	2300	250	45	154.90	CYD
ZH 1276	17.4	0.7	24.86	3900	670	240	15	173.45	CYD
ZH 1277	23.8	0.7	34.00	1400	600	330	220	181.90	CYD
ZH 1278	15.3	6.8	2.25	2400	320	4300	60	188.13	CYD
ZH 1279	18.1	8.9	2.03	8900	1700	260	120	224.70	BGN
ZH 1280	32.8	0.9	36.44	3600	1150	330	25	241.40	BGN
ZH 1281	30.0	1.2	25.50	5100	1150	960	90	257.10	BGN
ZH 1282	22.0	9.5	2.32	2700	2000	5000	160	272.45	BGN
ZH 1283	25.4	0.8	31.75	1500	1500	320	80	304.90	KNW
ZH 1284	20.0	5.3	3.77	2900	1900	230	15	330.35	KNW
ZH 1285	4.7	0.8	5.87	120	550	<50	350	347.25	KNW
ZH 1286	11.0	5.8	1.90	780	2600	<50	130	379.40	KNW

Metal contents expressed as percentages or parts per million (ppm) by weight of the samples as received. Analyst: J. L. Chapman.

LISC = Lithostratigraphic code: BGN = Ballagan Formation, CYD = Clyde Sandstone Formation, KNW = Kinnesswood Formation, SCK = Stockiemuir Sandstone Formation.

Samples from the Ballagan Formation are cementstones, the remainder are cornstones.
Note: in a pure dolomite, the Ca/Mg ratio would be 1.67.

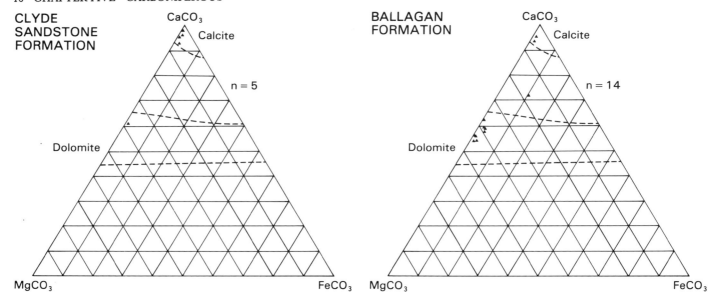

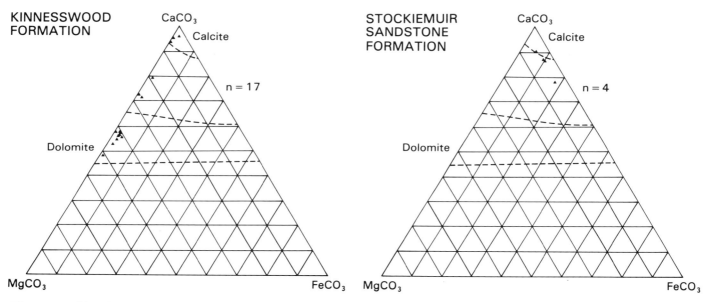

Figure 4 Chemical analyses of carbonate minerals in sediments of the Stratheden and Inverclyde groups, plotted by molar percent of CaCO₃, MgCO₃.

Data from Table 4. Field boundaries from Ford (1917).

A Clyde Sandstone Formation   B Ballagan Formation
C Kinnesswood Formation   D Stockiemuir Sandstone Formation

calcrete presents a fully mature profile in which laminar and pisolitic structures are developed and the carbonate is partially replaced by thin lenticular bodies of chert aligned parallel to the bedding. The calcrete bed, which was absent in the sequence cut by the Everton Borehole only 1.6 km distant, is overlain by an intraformational conglomerate composed largely of limestone clasts, some of which are cut by chert veins.   IBP, IHSH

At Auchengarth [190 645], where cross-bedding in the sandstones indicates deposition by southward-flowing rivers,

a 20 m sequence consisting mainly of red-brown and purple-grey silty mudstone occurs in the lower part of the Kinnesswood Formation. A number of thin, strongly cross-bedded white sandstone beds, with clasts of limestone or quartz, are intercalated in the mudstones. In addition to calcrete horizons, pale grey calcite mudstone (cementstone) occurs as lines of nodules and tabular beds up to 0.4 m thick. The sequence appears to be intermediate in character between the Kinnesswood and Ballagan formations.

Plate 5 Pedogenic carbonate (cornstone) in sandstone. Kinnesswood Formation, Inverkip [2016 7196] (D 2596).

In addition to the Auchengarth and Inverkip localities, there are good exposures of the Kinnesswood Formation in the area south of the River Clyde in the burn at Outerwards [234 661], in Martin Glen [232 669] and in the Shiel Water downstream from Coralees Bridge. The sharp basal contact of the formation with Stratheden Group sandstones and conglomerates is visible east of the bathing pool on the shore at Gourock [2379 7768]. The full thickness of the formation was 76 m in the Everton Borehole, but the Knocknairshill Borehole penetrated 104.5 m of cornstone-bearing sandstones and mudstones without reaching a base. In both boreholes, the junction of the Kinnesswood Formation with the overlying Ballagan Formation was transitional.   IBP

North of the River Clyde, the Kinnesswood Formation is generally thicker and more arenaceous than to the south. Mudstone and cornstone horizons are less commonly preserved but clasts of both are present throughout the sequence. Small quartz pebbles are scattered throughout the lower half of the formation and a few pebbles of other Highland rocks occur near the base. The sandstones may contain calcareous concretions or may have a partly calcareous matrix, especially in the lower part of the sequence. Only a few well-developed cornstone beds are present but some of these are traceable for distances of up to 0.5 km. The formation was 170 m thick in the Kipperoch Borehole. The best exposures are near Carman Reservoir [378 787] and near Wester Ardoch [353 768].   IHSH

Ballagan Formation (BGN)

Typically, the Ballagan Formation (Plate 6) consists of grey, silty, poorly laminated mudstones with numerous thin beds of limestone (cementstone). The cementstones, which are usually dolomitic according to analyses carried out on samples from the Kipperoch, Barnhill, Everton and Knocknairshill boreholes (Table 4 and Figure 4), occur as nodules and tabular beds, in a few cases showing internal lamination. Though rarely exceeding 0.2 m in thickness, individual beds may persist laterally for distances over 2 km. Gypsum, as veins and nodules, is common, traces of copper

Plate 6 Siltstone and mudstone with bands of fine-grained limestone or dolomite (cementstone). Ballagan Formation, Loch Thom [2501 7073] (D 3298).

and lead minerals occur in a few cementstones, and pseudomorphs after halite have been found in the mudstones at a number of localities. At the base of the formation, sandstone beds are intercalated among the mudstones in a zone, up to 20 m thick, which is transitional with the underlying Kinnesswood Formation.　　　　　　　　　　IBP, IHSH

South of the River Clyde, the best exposures are in the Beatock Burn [213 717], upstream from Everton. The Everton Borehole proved more than 50 m of typical grey silty mudstones and cementstones, the total thickness of the formation in the area being no more than 75 m. A similar thickness (70.5 m) was proved near Port Glasgow, in the Knocknairshill Borehole. At both localities, beds of grey and greenish grey sandstone were present in a transition zone at the base of the formation. On the hillslopes above Greenock, strata of the Ballagan Formation were formerly exposed in a number of stream sections now lost through expansion of the built-up area. Mudstones with cementstone beds are still visible, however, in a stream [285 754] north of Ingleston Street and at Crescent Street [286 752]. Strata of the forma-

tion were proved by commercial boreholes at Belville Street in Greenock.　　　　　　　　　　　　　　　IBP

The Ballagan Formation is best-developed in the Dumbarton area where it is estimated to be 140 m thick. There are good exposures in Murroch Glen [415 785], in Overtoun Burn [417 760] and around Whiteleys [380 768]. The Barnhill Borehole cut 128 m of strata in the uppermost part of the formation before encountering a fault. The Kipperoch Borehole penetrated 43 m of strata in the lower part of the formation, the sandy transition zone with the Kinnesswood Formation being about 20 m thick.　　　　　　　　IHSH

Clyde Sandstone Formation (CYD)

The strata assigned to the Clyde Sandstone Formation are thickest in the Gourock–Leap Moor area, where they occupy the axial zone of the north-north-east-trending Leap Moor Syncline, a structure which is considered to have formed at the end of Inverclyde Group times. Three subdivisions can locally be recognised. In ascending order they are the Knocknairshill Member (KKS), the Gourock Sandstone Member (GKA) and the Broadlee Glen Sandstone Member (BRLG). North of the River Clyde, the Clyde Sandstone Formation is represented by the Overtoun Sandstone Member (OVS). The Knocknairshill Member is made up of cornstone-bearing, upward-fining fluvial cycles indistinguishable from those in the Kinnesswood Formation. At its type section in the Knocknairshill Borehole, about 132 m of strata belonging to the member were proved, resting with a transitional junction on the Ballagan Formation. There are good exposures of the cornstone-bearing sandstones along the coast between Gourock and Cloch Point. Cross-bedding measurements obtained in this area by Bluck (1978, fig. 12), on rocks wrongly attributed to the upper part of the Upper Old Red Sandstone, indicate deposition by rivers which flowed generally towards the east-south-east.

The Gourock Sandstone Member is broadly similar but is distinguished by the presence of thick multistorey bodies of white pebbly sandstone. There are good exposures in quarries on Gourock Golf Course [226 764] and on the strongly featured hillslopes to the south.

The Broadlee Glen Sandstone Member crops out mainly in the northern part of Leapmoor Plantation where it consists of white and cream-coloured, trough cross-stratified sandstone with thin lenticular beds of grey siltstone and abundant coalified wood fragments. The best section is in Broadlee Glen [221 715 to 226 714]. White sandstones with coaly streaks, exposed on the coast [2015 7498] to the north of Lunderston Bay, probably belong to this member. They are thought to lie in a fault-slice. IBP

North of the River Clyde the formation is represented only by the Overtoun Sandstone Member. The complete sequence was penetrated by the Barnhill Borehole where it was 57 m thick. The lower part is dominated by grey, fine-grained, thin-bedded or partly concretionary sandstones. There are also some mudstones which contain bands or concretions of limestone, and erosive-based coarser sandstones with carbonate clasts. The upper part comprises largely sandstones with erosional bases and carbonate clasts; concretionary sandstones and mudstones are uncommon. Exposures occur in the Overtoun Burn [423 760]. IHSH

STRATHCLYDE GROUP

The Strathclyde Group (Paterson and Hall, 1986) comprises the rocks lying between the base of the thick sequence of lavas and associated volcaniclastic sediments known as the Clyde Plateau Volcanic Formation and the base of the Hurlet Limestone. Three divisions are recognised, the Clyde Plateau Volcanic Formation, the Kirkwood Formation and the Lawmuir Formation (Table 3). Where the Lawmuir Formation is absent, as in the attenuated sequence around Lochwinnoch, the top of the Strathclyde Group is drawn at the top of the Kirkwood Formation. IBP, IHSH

Clyde Plateau Volcanic Formation

The Clyde Plateau Volcanic Formation is the product of the most extensive of a series of volcanic episodes that took place during the Lower Carboniferous in central Scotland. South of the River Clyde, the formation occupies the high ground of the Renfrewshire Hills, and also extends westwards from Loch Thom to the coast at Cloch Point. Lavas also form high ground in the Beith–Barrhead area to the east of Lochwinnoch. North of the River Clyde, the formation occurs only in a small outlier on Ben Bowie. The volcanic rocks consist mainly of lavas in the range alkali olivine-basalt–hawaiite–mugearite, with subordinate pyroclastic rocks and volcaniclastic sediments. More siliceous lava types were, however, extruded from a major eruptive centre in the Misty Law area. Because of earth movements and the erosion that took place towards the end of the Inverclyde Group period, the Clyde Plateau Volcanic Formation rests disconformably on strata ranging from the Clyde Sandstone Formation to the Stratheden Group.

The eruptions seem to have been entirely subaerial. Tropical weathering between eruptive phases is suggested by the presence of reddened flow tops, and in many cases red-brown lateritic boles are present. Pyroclastic deposits are rare in the main basaltic sequence. There is little evidence for the existence of basaltic central volcanoes and most of the lava flows were probably derived from fissure eruptions. Feeder dykes necessary for such eruptions are difficult to identify, but where the lower part of the lava pile is exposed in deep valleys, numerous dykes with a predomiant northeast trend occur (p.32). Within the district these are particularly numerous along the projected continuation of the Dumbarton–Fintry line, a north-east-trending zone of vents, plugs and associated proximal pyroclastic deposits which is well developed in the Campsie and Kilpatrick hills (Whyte and MacDonald, 1974). Dyke-swarms, vents and plugs are also seen on Great Cumbrae and south Bute, along a further projection of the line to the south-west. It is possible that the volcanicity was related to a deep-crustal fracture along this line inherited from the Caledonian Orogeny.

The lava pile is thought to be thickest in the northern part of the Renfrewshire Hills, where it may reach 1000 m, and to thin somewhat towards the south-east. The lavas to the east of Lochwinnoch, which are separated from the main outcrop by the major structure known as the Paisley Ruck, may be only about 300 m thick. It would seem that the thinning was achieved by attenuation of the lower part of the lava pile (p.26), possibly because the ground to the south-east was uplifted by early movements on the Ruck.

The precise age and duration of the volcanic episode cannot be determined because no fossils of diagnostic value have been found. The rare mudstone beds within the tuffs have yielded only carbonised plant fragments. Build-up of the lava pile was accompanied by regional subsidence, and after volcanic activity came to an end the volcanic formation was progressively buried beneath sedimentary rocks of late Dinantian age. The earliest of these contain a high proportion of volcanic detritus washed in from the lava-terrain and constitute the markedly diachronous Kirkwood Formation.

Petrology

The lavas are all members of a mildly alkaline, alkali-basalt series, represented by compositions throughout the range ankaramite–basalt–hawaiite–mugearite–benmoreite–trachyte–rhyolite (Figure 5), comparable with other Dinan-

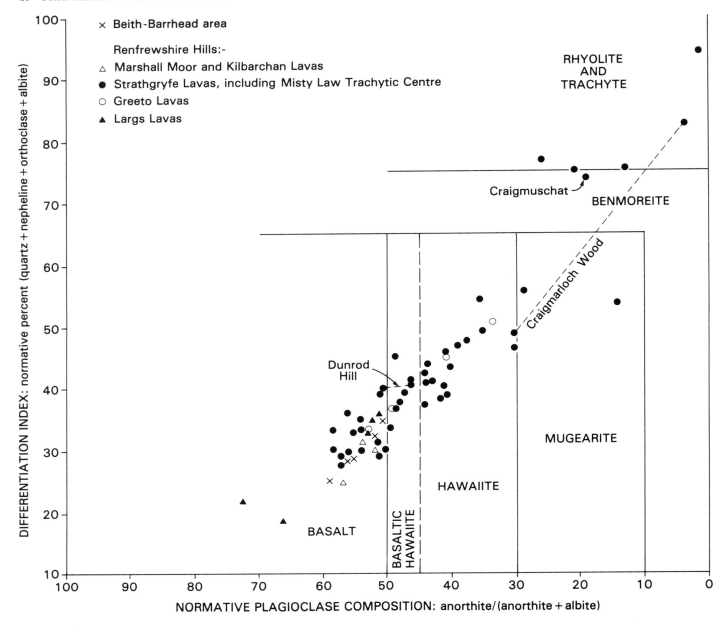

Figure 5 Range of composition of lavas in the Greenock district.

The classification, which is based upon CIPW norm calculations, is after Coombs and Wilkinson (1969) and Macdonald (1975).

Mugearites and more fractionated rocks are under-represented owing to the difficulty in obtaining fresh material of these compositions. Analyses from composite flows are linked by tie-lines. Data from Leitch and Scott (1917), Kennedy (1931; 1933), Smedley (1986) and unpublished BGS analyses by J G Fitton, University of Edinburgh.

tian lava assemblages of the Midland Valley (Macdonald, 1975). Analyses show that the most basic lavas are mainly hypersthene-normative, mildly alkaline or transitional rocks with a few of slightly silica-undersaturated, nepheline-normative composition. The more fractionated rocks are all hypersthene-quartz-normative. In terms of its alkali content, the assemblage is intermediate between a sodic alkali-basalt series, as typified by suites from Hawaii and the British Ter-

tiary Province, and a potassic alkali-basalt series, as represented by suites from Gough Island and Tristan da Cunha. There is a tendency for the more basic rocks to be mildly sodic and for compositions to become more potassic with increased fractionation, so that some lavas of intermediate composition may be termed trachybasalts and trachyandesites.

The lavas, particularly the more basic varieties, can be remarkably fresh, although olivine is usually replaced by red-brown pseudomorphs. Less fresh material shows varying degrees of albitisation, chloritisation, oxidation, hydration and replacement by carbonate. Albitisation in particular can lead to considerable difficulties in petrographic and petrochemical classification.

The more basic rocks are almost all porphyritic with various combinations of plagioclase (white), olivine (red-brown) and clinopyroxene (black) phenocrysts. On this basis they have been classified in the field and in thin sections into six main types (Table 5), following MacGregor (1928). Of these, the Dalmeny (B^D), Dunsapie (B^{Du}) and Hillhouse (B^H) types are all olivine-basalts, and the very mafic Craiglockhart type (B^{Ck}) could be termed ankaramite. The feldsparphyric lavas, allotted on a petrographical basis to the Markle and Jedburgh types, have compositions in the range olivine-basalt through basaltic hawaiite to hawaiite. Within this range, most of the basaltic Markle lavas as defined by MacGregor (1928) contain microphenocrysts of olivine visible in hand specimen, and are identified on the map by the symbol B^M. Such lavas occur in the Beith–Barrhead area and in the upper part of the Renfrewshire Hills succession. Other feldsparphyric lavas, notably in the central part of the Renfrewshire Hills succession, contain no olivine phenocrysts and have little or no visible olivine in the groundmass. Analyses indicate that the majority of these are basaltic hawaiites or hawaiites, although some may be classed as basalts. The flows of this group are referred to by the general term 'Markle lavas' and are indicated on the map by the combined symbol fBW which represents feldsparphyric lavas of basaltic to hawaiitic composition. A similar distinction is made between true 'Jedburgh basalts' (B^J) and 'Jedburgh lavas' of unspecified basaltic to hawaiitic composition (BW).

Nonporphyritic lavas in the district are usually somewhat paler than the basalts and typically show a pronounced platy jointing due to flow-alignment of feldspar crystals. In accordance with usual practice in the Scottish Dinantian volcanic sequences, these rocks have been classified as mugearites (W^M), although the available analyses show that they range from hawaiite through mugearite to benmoreite or trachyandesite in composition. In thin section they consist mainly of small fluxioned oligoclase feldspar laths with a relatively high proportion of iron-titanium oxide grains. Clinopyroxene and olivine are commonly absent but where present are less abundant than in the basalts and are usually altered. Some rocks contain scattered feldspar phenocrysts and are thus transitional to hawaiitic Markle or Jedburgh lavas, often within a single flow. Some flows, mainly associated with the Misty Law Trachytic Centre, contain phenocrysts of plagioclase, and in some cases clinopyroxene and/or amphibole, in a groundmass containing appreciable amounts of alkali-feldspar. Such rocks have been classed as trachybasalt (W) or trachyandesite (N) according to the composition of the plagioclase phenocrysts.

The trachytic and rhyolitic rocks are commonly grey-green when fresh, but usually weather to pink or purplish hues. At many outcrops the lavas have an irregular and undulating slabby jointing, probably reflecting primary viscous flow-lamination, but at others a brittle, flaky parting is developed. The rhyolites often display pronounced flow-banding in which quartz is visible, and spherulitic textures occur in places. Pink, rectangular phenocrysts of alkali-feldspar, commonly sanidine and/or anorthoclase but also orthoclase-perthite and albite, are usually present but fresh mafic silicates have not been observed, the only mafic phases being iron-titanium oxides and chlorite.

Table 5 Nomenclature of basic igneous rocks of Carboniferous and Permian age in the Midland Valley of Scotland

| Basalt type of MacGregor (1928) | Phenocrysts | | Chemical classification of Macdonald (1975) | Type locality |
	abundant	sometimes present in lesser amounts		
Macroporphyritic (phenocrysts >2 mm)				
Markle	pl	± ol, Fe	pl ± ol ± Fe-phyric basalts, basaltic hawaiites or hawaiites	Markle Quarry, East Lothian (flow)
Dunsapie	pl + ol + cpx	± Fe	Ol + cpx + pl ± Fe-phyric basaltic hawaiites or ol + cpx + pl-phyric basalts	Dunsapie Hill, Edinburgh (vent intrusion)
Craiglockhart	ol + cpx		Ankaramite	Craiglockhart Hill, Edinburgh (flow)
Microporphyritic (phenocrysts <2 mm)				
Jedburgh	pl	± ol, Fe	pl ± ol ± Fe-phyric basaltic hawaiites, hawaiites and in some cases basalt	Little Caldon, Stirlingshire (plug). Also in Jedburgh area
Dalmeny	ol	± cpx, pl	ol ± cpx-phyric basalt	Dalmeny Church, West Lothian (flow)
Hillhouse	ol + cpx		ol ± cpx-phyric basalt (rarely basanite)	Hillhouse Quarry, West Lothian (sill)

pl = plagioclase, ol = olivine, cpx = clinopyroxene, Fe = iron-titanium oxides

Composite flows

A relatively common feature of the mugearite/Markle lava sequence of the Strathgryfe Lavas (p.23) is the presence of composite flows. In these, the basal part usually consists of nonporphyritic lava of mugearitic appearance, which grades upwards into a feldsparphyric rock of hawaiitic Markle type. Good examples have been described by Kennedy (1931) from Dunrod Hill, Inverkip, but others occur throughout the area, for example at Muirfauldhouse [337 601]. In most cases the nonporphyritic lava is almost identical petrographically to the groundmass of the porphyritic variety and analyses from the Dunrod flow (Kennedy, 1931) show that the two lava types are chemically similar (Figure 5). Any slight differences are attributable to the presence or absence of labradorite phenocrysts. The gradational transition between the lava types clearly indicates that each flow is the product of a single eruption. It is possible that differentiation may have occurred in situ through a combination of flotation and flow differentiation, although Kennedy preferred a differential flow mechanism whereby earlier pulses, free of suspended phenocrysts, are subsequently overlapped by faster-moving later pulses carrying the phenocrysts. Whatever the eruptive mechanism, these flows provide clear evidence of the intimate petrogenetic relationship between the basaltic hawaiite, hawaiite and mugearite lavas, and demonstrate the effectiveness of crystal fractionation processes in the evolution of this series.

A composite flow at Craigmarloch Wood [345 719], northwest of Kilmacolm, is of a distinctly different type. It consists of two phases of widely differing composition, a 'basic mugearite' and a feldsparphyric trachyandesite (Figure 5; Kennedy, 1933). The two phases are intermixed in an irregular manner, although the trachyandesite tends to occupy a central position and contains xenoliths and schlieren of mugearite in a wide transition zone. The evidence in this instance suggests the mixing of two distinct magmas, erupted simultaneously, but probably from different effusive centres.

Sequence in the Renfrewshire Hills

A major petrographical division of the lavas of the Renfrewshire Hills and adjoining areas into Upper and Lower groups, proposed by Richey (1928), was later elaborated by Johnstone (1965) into a stratigraphical classification with five subdivisions, applicable to the Renfrewshire Hills only. The present classification (Table 6), which is modified from that of Johnstone (1965), consists of six units of member status. Their distribution is illustrated in simplified form in Figure 6.

NODDSDALE VOLCANICLASTIC BEDS   Purplish tuffs, agglomerates and volcaniclastic sediments at the base of the volcanic formation in the area between Loch Thom and Largs represent an early explosive phase prior to extrusion of the main lava pile. The complete unit was penetrated by the Largs Borehole, which proved the following sequence:

	Thickness	*Depth*
	m	m
Lava, grey, feldsparphyric	107.36 +	107.36
Volcanic detritus, fine- to coarse-grained with lava blocks up to 0.2 m long, bedded in places. Fine-grained, with accretionary lapilli, in basal 3 m	43.86	151.22
Mudstone, silty, purple-brown, mainly massive	11.16	162.38
Sandstone, tuffaceous, grey-purple and purple-brown with clasts of lava and quartz	7.07	169.45

In this area, east of the Largs Fault-zone, the deposit rests upon cornstone-bearing sandstones of the Kinnesswood Formation.

Tuffs and agglomerates containing blocks, up to 0.3 m long, of locally derived basalt and, in a few cases, sandstone are exposed in tributaries of the Noddsdale Water between Largs and the Muirshiel Fault. In the Noddsdale and Bailieland burns and on the hillside [218 608] 300 m east of Kilburn Farm, the deposits are characterised by the presence of quartz pebbles, presumably derived from conglomerate beds. The finer-grained tuffs are commonly well-bedded and grade into water-laid volcaniclastic and quartzose sandstones, some of which contain carbonised plant debris, for example in the Noddsdale Burn [213 615].   DS

Farther north, in the area west of the Largs Fault-zone between Outerwards Reservoir and Routen Bridge, the Noddsdale Volcaniclastic Beds are up to 200 m thick and rest upon strata high in the Clyde Sandstone Formation. The deposit, here composed of purple-grey tuff and agglomerate, is coarse-grained in places with lava blocks up to 0.3 m in length. Bedding, with steep inclinations that may be partly due to original depositional dips, is visible at [250 686] but more usually the tuffs and agglomerates are unbedded.

At exposures in the Rotten Burn [250 703], south of Loch Thom, the deposit is represented by some 15 m of purple-brown mudstone, lying between fractures in the Largs Fault-zone.   IBP

LARGS LAVAS   The lowest lavas in the Clyde Plateau Volcanic Formation form the major scarp on the east side of Noddsdale, south of the Muirshiel Fault, and can be traced northwards in the steep belt adjacent to the Largs Fault-zone on the east side of Knockencorsan Hill and in the Rotten Burn. They consist of up to 130 m of relatively uniform microporphyritic basalts of Dalmeny type with a few flows of Hillhouse basalt. Flows of Craiglockhart basalt (ankaramite) are intercalated in the upper part of the sequence between Harplaw and the Tourgill Burn.

GREETO LAVAS   The lavas of this unit are well exposed in the deep valleys of the Greeto and Gogo waters close to their confluence [230 593]. They also crop out extensively on the upper part of the main scarp on the east side of Noddsdale as far north as the Muirshiel Fault. The unit is up to 110 m thick and comprises a number of flows of basaltic or hawaiitic Jedburgh lava, together with a few flows of Markle lava. The Jedburgh lavas are characteristically fresh and massive in their central parts and can form cliff-features up to 12 m high, as at Cauld Rocks [220 602]. This is in marked contrast to the lower, more rounded features formed by the more basic flows in the underlying Largs Lavas.

Table 6 Divisions of the Clyde Plateau Volcanic Formation in the Renfrewshire Hills

Member	Petrography	Distribution	Maximum thickness (metres)	Richey (1928)	Johnstone (1965)
Kilbarchan Lavas	Mainly microporphyritic basalts of Dalmeny type; a few flows of Hillhouse basalt and Craiglockhart ankaramite	Eastern edge of hills from Lochwinnoch to Langbank	75	Upper Group	5
Marshall Moor Lavas	Macroporphyritic basalts of Dunsapie type and ankaramites of Craiglockhart type; some Markle basalts with minor olivine phenocrysts	Eastern edge of hills from Lochwinnoch to Langbank	75	Lower Group	4
Strathgryfe Lavas	Flows of Markle lava (without olivine microphenocrysts and probably mostly of hawaiitic composition) and mugearite in approximately equal proportions; some Jedburgh lavas (basalt and hawaiite); up to three groups of Craiglockhart ankaramites and Dunsapie basalts in upper part. MISTY LAW TRACHYTIC CENTRE: trachytes and rhyolites with associated tuffs and agglomerates; minor trachyandesite and trachybasalt; one major basaltic intercalation	Northern and central part of the hills from Greenock to the Muirshiel Fault, member not divided. South of the Muirshiel Fault the member is divided into: (i) upper Strathgryfe Lavas of the River Calder valley and Ladyland Moor (max. 500 m thick), (ii) Misty Law Trachytic Centre around Misty Law, Hill of Stake and Irish Law (max. 300 m thick), (iii) lower Strathgryfe Lavas of the Greeto Water valley, Rowantree Hill and Peak Hill (max. 250 m thick)	750		3
Greeto Lavas	Microporphyritic basalts and hawaiites of Jedburgh type with a few of Markle type	Top of scarp slope, east of Noddsdale Water from Largs to Muirshiel Fault; valley of the Greeto Water	110		2
Largs Lavas	Mainly microporphyritic basalts of Dalmeny type with some macroporphyritic Craiglockhart ankaramites	Scarp slope east of Noddsdale Water from Largs to Muirshiel Fault; Knockencorsan Hill	130		
Noddsdale Volcaniclastic Beds	Mainly basaltic tuffs and agglomerates with some derived, volcaniclastic, bedded sandstones	Base of scarp slope east of Noddsdale Water; Noddsdale Water; Knockencorsan Hill; Rotten Burn	0–200		1

STRATHGRYFE LAVAS   In the area north of the Muirshiel Fault, a thick sequence composed mainly of mugearites and Markle lavas, in approximately equal amounts, is assigned to the Strathgryfe Lavas. South of the Muirshiel Fault a unit of trachytic and rhyolitic rocks, here termed the Misty Law Trachytic Centre, divides the Strathgryfe Lavas into an upper and a lower part (Figure 6).

Few of the Markle lavas contain olivine phenocrysts and the groundmass is also relatively felsic. It seems likely, therefore, that they are mostly of hawaiitic rather than basaltic composition. Textures are variably porphyritic and sparsely porphyritic varieties can in places be observed to grade into typical mugearite within a single flow, as at exposures north of Muirfauldhouse [337 601].   DS

In addition to the mugearites and olivine-free Markle lavas, there is also, near the top of the unit, a thin development of ankaramitic lavas and olivine-bearing Markle basalts which crop out around Knaps Loch and to the south and east of Langbank. There are also a few flows of microporphyritic hawaiitic or basaltic Jedburgh lava

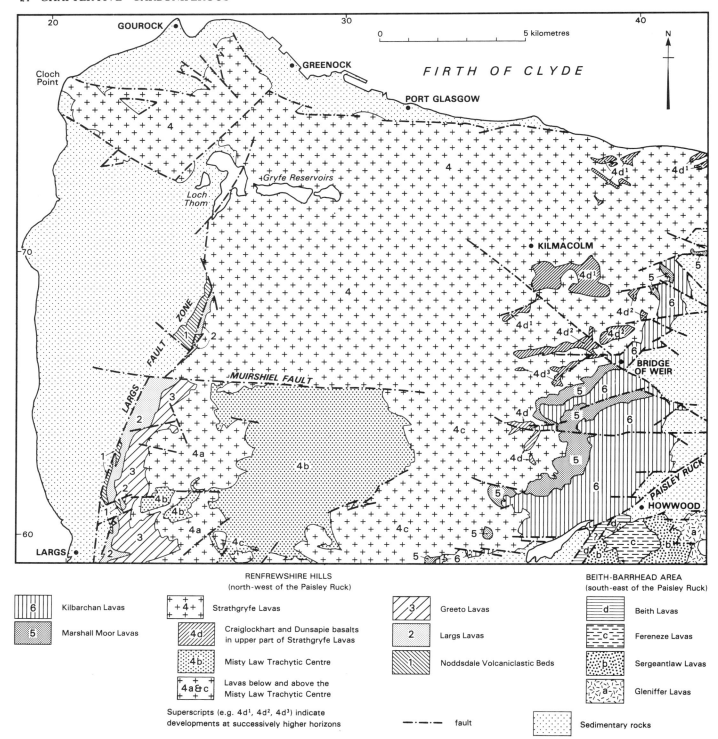

Figure 6 Stratigraphical divisions of the Clyde Plateau Volcanic Formation in the Renfrewshire Hills (Table 5) and the Barrhead–Beith area (Table 6)

throughout the unit. Trachyandesite occurs within the composite flow at Craigmarloch Wood [345 719] (p.22) and south and east of Cauldside [323 705].

From Cloch Point and Dunrod Hill to Port Glasgow, the Strathgryfe Lavas rest directly upon strata at various levels within the Clyde Sandstone Formation, with no intervening

lower lava unit. However, in the Knocknairshill Borehole the lavas were separated from the sandstones by 40 m of volcaniclastic sediments. Brown blocky mudstone, probably equivalent to part of these, is exposed beneath a series of hawaiitic flows at Wallace's Loop [309 740]. IBP

A good impression of the relationship between flows is

given by boreholes drilled in the Calder Valley around Clovenstone [333 616], together with good natural exposures in the immediate vicinity. Flows of mugearite and Markle lava occur in approximately equal numbers. Individual flows are between 3 m and 20 m thick, with an average of about 10 m, and up to four flows of similar rock type commonly occur in succession. In the boreholes, the flows are almost invariably separated by boles, up to one metre thick, which are rarely seen at outcrop. Most flows contain bands of fresh, massive lava, amygdaloidal lava and hydrothermally altered, autobrecciated, slaggy material. The last is commonly soft, friable and in some cases makes up 90 per cent of the flow. Natural exposures consist mainly of the fresh and massive centres of flows. Even within the area covered by the boreholes, individual flows can seldom be traced for more than 1 km, thicknesses vary considerably and local overlaps and unconformities are common.

The width of the Strathgryfe Lavas outcrop in the north is in part a reflection of the great thickness of the sequence, which is estimated to be up to 750 m, but is also due to the gentle south-eastward dip which gives rise to extensive dip slopes, as near Westside [335 695] and Barscube Hill [359 715]. There may also be repetition by faulting. All the rock types form good trap features in places, the mugearites in particular forming cliffs up to 15 m high, as at Craig Minnan [322 641]. The Markle lavas tend to form lower scarp features and knobbly outcrops on dip slopes.

MISTY LAW TRACHYTIC CENTRE   Trachytic and rhyolitic lavas, pyroclastic rocks and associated intrusions, forming an outcrop up to 8 km wide in the area around Misty Law [295 620], were originally thought to lie within a collapsed caldera and thus to be of later date than the surrounding basaltic lavas. However, Johnstone (1965) concluded that they form a shallow cone, the remnant of a central volcano, intercalated near the middle of the Strathgryfe Lavas.

Six main phases of extrusive activity have been recognised. In order of formation the products of these are:

i   The trachyte lavas exposed in streams south-east of Waterhead Moor [263 621] and around the headwaters of the River Calder. Bodies of trachyte lava, formerly regarded as sills, which flank the Greeto Water at Girtley Hill [232 614] and west of Brown Hill [243 605], are intercalated in the lower Strathgryfe Lavas and appear to be precursors of the main extrusive phase.

ii   The pyroclastic rocks in the major scarp east of Waterhead Moor [269 264], in the headwaters of the Gogo Water and Routdane Burn, in Murchan Burn and in Surge Burn. These are considered to be distal parts of a pyroclastic apron deposited around the trachytic vents of Little Irish Law [261 594] and Knockside Hills (Sheet 22W). In Murchan Burn [290 609] the tuffs contain lenses of carbonaceous mudstone with carbonised plant fragments.

iii   The trachytic lavas and welded tuffs forming high ground which extends northwards and north-eastwards from the Gogo Water headwaters across the whole of the centre.

iv   The basaltic and intermediate lavas forming outliers on the Tongue [280 610] and more continuous outcrops around East Girt Hill [279 627], Burnt Hill [270 634] and Queenside Loch [292 642]. Basic rocks on Irish Law, formerly interpreted as plugs, may also be lava outliers.

v   The massive flows of trachyte and rhyolite overlying the basic lavas on and west of Queenside Hill [293 638], on Hill of Stake [274 630] and extending south-eastwards from East Girt Hill [279 627] to Little Misty Law [299 621].

vi   The flows of trachyandesite and trachybasalt capping Totterie Law [293 621] and Misty Law and also occurring close to the south-eastern edge of the centre at [313 623], [313 618] and [305 613]. Some of these bodies were previously mapped as plugs.

The intrusive rocks of the centre, consisting of vent agglomerate, trachytic plugs and dykes, are described in the next chapter.

Along the western margin of the centre, the contact of the basal trachytes with the Strathgryfe Lavas is not exposed but appears to be conformable and to dip gently to the south-east. The more steeply dipping upper surface of the cone is believed to be overstepped by the upper Strathgryfe Lavas but, where exposed, the contacts are generally faulted. Unfaulted junctions are seen only in the River Garnock [293 593] and in the Rough Burn [315 617] (Johnstone, 1965), outwith the faulted south-east margin of the main mass. The distal parts of the cone, where it interdigitates with more basic rocks, are mainly obscured by faulting, for example by the Muirshiel Fault (Figure 4), but may be represented by thin developments of trachyte and trachytic tuff at South Burnt Hill [255 651], at Feuside Hill [252 597] and also at a locality [250 612] some 1500 m north of Feuside Hill.

MARSHALL MOOR LAVAS AND KILBARCHAN LAVAS   The lavas in the highest two divisions of the Renfrewshire Hills sequence are considerably more basic than those in the underlying units. The change is abrupt and implies a significant time interval between two petrologically distinct volcanic episodes.

The Marshall Moor Lavas consist mainly of Dunsapie basalts, with olivine-bearing Markle basalts and a few ankaramite (Craiglockhart) flows. Hawaiitic Markle lavas and mugearites are conspicuously absent. The unit can be traced from outliers around Lochwinnoch Golf Course northwards, through its thickest development in the area between Peockstone [357 610] and Bridge of Weir, to discontinuous, faulted outcrops of a few flows between Yonderton [390 667] and Barochan [405 695].

The Kilbarchan Lavas consist of hard, fresh, blue, microporphyritic Dalmeny basalts with rare flows of Hillhouse type. A few isolated flows of ankaramitic Craiglockhart type occur around Castle Semple and at Gladstone. The unit crops out over a wide area, forming extensive dip slopes in the area between Castle Semple Loch and Ranfurly and extends northwards, somewhat thinned, to Barochan where the outcrop is terminated by a system of major ENE- and WNW-trending faults.

In the south, outcrops of both the Marshall Moor and Kilbarchan lavas are terminated by a major east–west fault system at Lochwinnoch, but thin outliers occur at the top of the lava succession between Glenlora and the Maich Water [324 587].

Lavas of the Beith – Barrhead area

Lavas in the extreme south-east of the district are part of an outcrop, about 16 km long by 6 km wide, which occupies high ground between Beith and Barrhead. The lavas, consisting almost entirely of olivine-basalts, are disposed in a shallow north-east-trending anticline which plunges at a low angle towards the south-west. The northern margin of the outcrop, in the present district and in the adjoining area to the east, is a complicated fault-structure, to the north of which is low ground occupied by sedimentary rocks younger than the lavas. Within the present district, the lavas dip towards the north-west at 8° or less. Their overall thickness may be no more than 250 to 300 m.

Four divisions of the lavas, recognised in the district to the south (Richey et al., 1930), can be traced into the present district and the adjoining area to the east (Hall and Forsyth, in preparation). The divisions, which are considered to have the status of members, have been renamed to accord with current ideas of stratigraphical nomenclature (Table 7).

The oldest flows in the area, which crop out to the north of Whittliemuir Midton Loch [417 592], are assigned to the Gleniffer Lavas. They form a distinctive group of Markle basalts, characterised by flows with abundant, platy phenocrysts of plagioclase up to 25 mm long and microphenocrysts of olivine. Overlying these to the west are the Sergeantlaw Lavas. These consist of mafic olivine-basalts of Dunsapie and Dalmeny type, which form prominent scarp features around Walls Hill [412 589]. Mugearites occur locally near the base. South of West Gavin [380 589], Dalmeny basalts of the Sergeantlaw Lavas appear to rest upon basaltic tuff and agglomerate and it is possible that in this area the Gleniffer Lavas have been overlapped. Resting upon the Sergeantlaw Lavas, and forming a broad dip slope outcrop south of Muirdykes [395 591], are several flows of Markle and Dunsapie basalt which are assigned to the Fereneze Lavas. Poorly exposed Dalmeny basalts, which appear to be overlain by volcaniclastic sediments of the Kirkwood Formation in downfaulted blocks, occupy low ground between West Gavin [380 590] and the Black Cart Water. Their stratigraphical position is uncertain, but they are tentatively assigned to the Beith Lavas.

Correlation with the lavas of the Renfrewshire Hills is difficult, as the two outcrops are separated from each other by the continuation of the Paisley Ruck. However, the almost exclusively olivine-basalt composition of the lavas in the Beith – Barrhead area, and the presence of many flows of the mafic Dunsapie and Dalmeny varieties, suggests a correlation with the highest parts of the Renfrewshire Hills sequence, the Marshall Moor and Kilbarchan lavas. Outcrops

and borehole occurrences of lava within the intervening low-lying land are also of Dalmeny basalt. Some 6 km to the east of the present district, the Glenburn Borehole showed that the basal lavas of the Beith – Barrhead sequence rest directly upon the Kinnesswood Formation. Hence it appears that the lower lava groups of the Renfrewshire Hills die out to the south-east, across the Paisley Ruck. The Beith – Barrhead succession includes a greater variety of lava types than is seen in the Marshall Moor and Kilbarchan lavas, however, and may include lavas younger than any represented in the Renfrewshire Hills.   DS

Volcanic rocks of Ben Bowie

North of the River Clyde, a sequence of volcanic rocks, 60 m thick, occurs in an outlier on the summit of Ben Bowie, 4 km east of Helensburgh. A lower unit, exposed only in the south-west of the outlier, consists of 15 m of basal tuffs overlain by a 6 m thick flow of Jedburgh lava. This unit is overstepped by a sequence of Markle lavas with two flows of mugearite which could be an attenuated, distal part of either the Renfrewshire Hills or the Kilpatrick Hills lava fields. However, the occurrence of the basal pyroclastic unit, which is cut by a small plug of Markle basalt in the extreme south-west, and the rubbly nature of some of the flows suggest that they belong to a proximal sequence, extruded from a local centre.   IHSH

Kirkwood Formation (KRW)

The Kirkwood Formation consists of a highly variable thickness of detrital material derived by erosion from the lavas of the Clyde Plateau Volcanic Formation. It rests on the uneven weathered surface of the latter and is notably diachronous, interdigitating with the Lawmuir and Lugton Limestone formations at levels ranging from below the Newton and Castlehead coals east of Howwood to below the Dockra Limestone west of Lochwinnoch (Figure 7). It has not been found north of Kilbarchan. The detritus ranges in grade from mud to cobbles of basaltic lava. The finer sediments are mostly red or red-brown in colour but some show yellow, green or purple tints. In the Lora Burn Borehole, just to the south of the district, the formation was represented by 5.6 m of greenish grey and brown mottled mudstones with scattered plant fragments.   SKM

Lawmuir Formation (LWM)

The Lawmuir Formation includes the strata above the Clyde Plateau Volcanic and Kirkwood formations and below the Hurlet Limestone. They were formerly included in the Calciferous Sandstone Measures. The formation occurs in a belt up to 3 km wide along the south-eastern margin of the district from north of Houston to Howwood and varies in thickness from less than 20 m to 150 m; it is absent farther to the south-west, around Lochwinnoch, where the time-equivalent strata are probably included in the Kirkwood Formation (Figure 7). It consists of a heterogeneous assemblage of sedimentary rocks, including sandstones, siltstones, mudstones, seatrocks and coals, together with a bed of basaltic tuff. In the upper part of the formation, there are a number of marine mudstones and limestones with rich

Table 7 Divisions of the Beith – Barrhead lava succession

New names	Names used by Richey et al. (1930)
Beith Lavas	Upper Group
Fereneze Lavas	Lower Group (c)
Sergeantlaw Lavas	Lower Group (b)
Gleniffer Lavas	Lower Group (a)

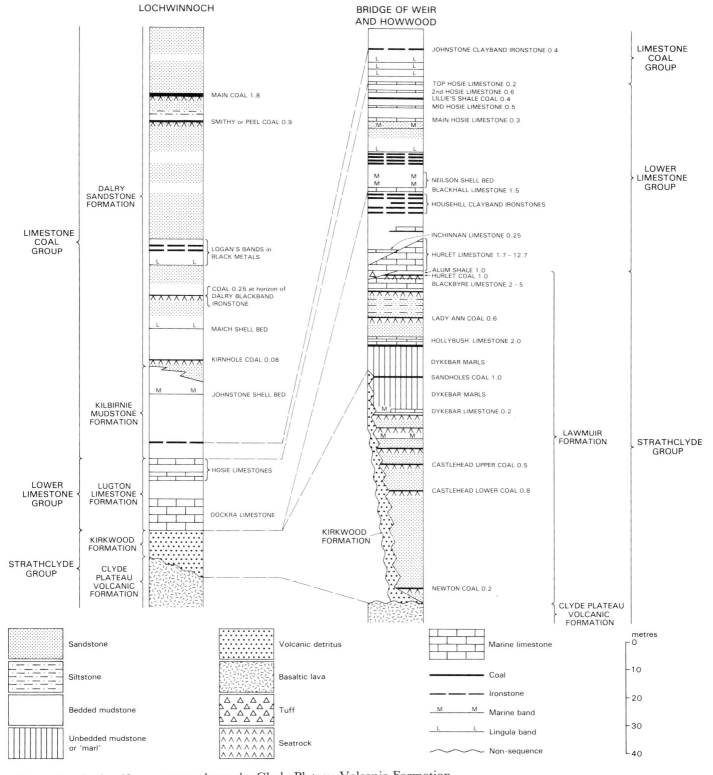

Figure 7 Carboniferous strata above the Clyde Plateau Volcanic Formation

and varied faunas. Details are given by Hinxman et al. (1920, pp.5–28) and Forsyth (1978, pp.1–8).

The sequence in the lower part of the Lawmuir Formation shows marked lateral variation, but in the upper part the strata are organised in generally upward-coarsening cycles.

Newton Coal to Dykebar Limestone

This part of the sequence is exceedingly variable in thickness, ranging from 0 to 70 m. The Newton Coal is the lowest of the seams which, at Johnstone, just east of the district, combine to form the Quarrelton Thick Coal (Hinxman et al., 1920, pp.19–28), the others being the Castlehead Lower and Upper coals. In this district, the three coals can be recognised as separate seams only at the southern end of the outcrop, where the intervening sediments consist mainly of sandstone and seatrock, with some volcanic detritus. They are absent under the Howwood Syncline but form a diminished representative of the Quarrelton Thick Coal around Kilbarchan. This seam is up to 2 m thick, including seatclay partings, where it was mined on Barr Hill [408 636], but is not known to the north. The strata above the Castlehead Upper Coal are mostly sandstones but include three variable coals. A marine band that occurs in a thin mudstone about 10 m below the Dykebar Limestone is the lowest fossiliferous horizon above the Clyde Plateau Volcanic Formation. North of Crosslee little is known of the sequence except that sandstones are locally prominently exposed.

Dykebar Limestone to Hollybush Limestone

The Dykebar Limestone is generally represented by a massive calcareous mudstone with a marine fauna of bivalves and brachiopods including, in particular, *Spirifer* cf. *crassus*. It lies about 30 m below the Hollybush Limestone. The intervening strata consist largely of the Dykebar Marls—a sequence of poorly bedded, variegated mudstones locally containing clusters of sphaerosiderite grains and with roots in places. Sandstones also occur, particularly in the lower part. The Sandholes Coal is only recognisable north of Kilbarchan. It is up to 1.15 m thick in several leaves, lies 12 m to 20 m below the Hollybush Limestone and was mined to some extent at its type locality [412 643].

Hollybush Limestone to Blackbyre Limestone

The Hollybush Limestone consists of two or more beds of Limestone within several metres of calcareous mudstone that are visible in two old quarries [412 666; 411 606] and in the River Gryfe [399 657; 415 664; 416 663]. They contain the lowest rich marine fauna in the Carboniferous of western Scotland. This fauna includes corals (mainly *Lithostrotion*), the echinoid *Archaeocidaris*, molluscs and brachiopods, among which *Latiproductus* cf. *latissimus* is characteristic. Between the marine beds and the variable Lady Ann Coal, which in places in 0.66 m thick, are 10 m of sandstone and 2 m of seatrock. Above the coal, in the Howwood Syncline, there is a band of basaltic tuff.

Blackbyre Limestone

The Blackbyre Limestone consists of 2 to 5 m of limestone with bands of calcareous mudstone. It lies close below the Hurlet Coal, and locally has been leached and affected by the roots of the coal-forming plants. The result is a white nodular limestone with a matrix of green clay, which passes in places into a clayrock with limestone nodules and bands. It is exposed at low water in the River Gryfe [404 656] and in a small stream [411 607] near Howwood. Carruthers (*in* Hinxman et al., 1920, pp.17–18) misidentified these exposures as the White Limestone and the westernmost exposure of the Hollybush Limestone in the River Gryfe as the Blackbyre Limestone (see Forsyth and Wilson, 1965, pp.68–69; Forsyth, 1978, pp.4–5).   IHF

The Blackbyre Limestone contains a rich marine fauna including corals such as *Diphyphyllum* and brachiopods such as *Echinoconchus* cf. *punctatus* and *Pleuropugnoides*. Up to 5 m of seatclay separate it from the Hurlet Coal.   PJB

Hurlet Coal and Alum Shale

The Hurlet Coal is a low-quality, pyritous seam up to 1.5 m thick, locally in leaves. It was mined in the Howwood Syncline and east of Bridge of Weir but there is little information about most of the workings. The mudstone between it and the Hurlet Limestone is known as the Alum Shale because it was exploited for alum at Lennoxtown, north of Glasgow. It reaches over a metre in thickness in places and has a locally rich fauna of brachiopods and marine bivalves. A layer of basaltic tuff and agglomerate which underlies the Hurlet Limestone in the Howwood area may replace the Hurlet Coal in places.

LOWER LIMESTONE GROUP (LLGS)

The Lower Limestone Group occurs only in the south-east of the district, between Bridge of Weir and Kilbarchan, in the Howwood Syncline, and at Lochwinnoch. The first two areas form part of the Central Coalfield, and nomenclature follows that of Hinxman et al. (1920). The Lochwinnoch area lies in the Dalry Basin of north Ayrshire, where a new nomenclature is now applied. Correlation between the two systems is shown in Figure 7.

Bridge of Weir, Kilbarchan and Howwood area

Information comes largely from boreholes, supplemented by stream sections in the River Gryfe and Locher Water and by quarries in the Hurlet Limestone in the Howwood Syncline. Details are given by Hinxman et al. (1920, pp.29–39). The base of the unit was formerly taken at the base of the Hurlet Coal but is now drawn at the base of the Hurlet Limestone; the top is at the top of the Top Hosie Limestone. The group is 90 m thick in the Howwood Syncline (Forsyth, 1978), where the highest 10 m are not preserved. The thickness in the Bridge of Weir–Kilbarchan area is about 60 m.

The sequence consists largely of dark grey bedded mudstones. Siltstones and silty sandstones are rare. The limestones which give the group its name are marine and up to seven in number, excluding the limy ironstones just below the Blackhall Limestone, which contain only ostracods and fish debris. Only two of the limestones are more than a metre thick, the remainder being generally in the range 0.25 to 0.6 m. The Blackhall Limestone is 0.45 to 1.9 m thick. The Hurlet Limestone, which varies in thickness from 1.7 to 12.7 m, consists of argillaceous shelly limestone with calcareous mudstone partings; to the east of the district, in the Paisley area, it is generally a crinoidal limestone less than 1 m thick (Forsyth and Wilson, 1965). Most of the limestones are composed largely of crinoid columnals but the Top Hosie Limestone is very argillaceous with scattered shell and crinoid debris. Clayband ironstones are scattered throughout the sequence, occurring in abundance just below the Blackhall Limestone (the Househill Clayband Ironstones) and midway between it and the Main Hosie Limestone. Two thin and impersistent coals, nowhere more than 0.35 m thick, occur below the Blackhall Limestone. A more persistent seam, the Lillie's Shale Coal, normally consists of 0.2 m of coal on 0.2 m of cannel or cannelly mudstone ('shale') but reaches 0.2 m of coal on 0.25 m of 'shale' in the Locher Water [402 648] and 0.4 m of coal on 0.45 m of 'shale' in a nearby borehole [421 658]. The seam was exploited for both coal and oil-shale but little is known about the long-abandoned workings.   IHF

The limestones and associated mudstones are marine and some, such as the Neilson Shell Bed above the Blackhall Limestone (Wilson, 1966), yield rich faunas. In the Shell Bed, which is exposed in the River Gryfe [404 657], *Crurithyris urii* and *Tornquistia youngi* are abundant. Most of the mudstones, however, are either barren or yield only ostracods, fish remains and *Curvirimula*. Such faunas occur particularly between the Hurlet Limestone and the Blackhall Limestone. The Hosie Sandstone, which most unusually for a Scottish Carboniferous sandstone contains a marine fauna, mainly of brachiopods, is represented by a bed of silty sandstone below the Main Hosie Limestone.   PJB

Lochwinnoch area

Only the upper part of the Lower Limestone Group, referred to the Lugton Limestone Formation, is present. The Dockra Limestone, which lies near the base in the Lochwinnoch area, is equivalent to the Blackhall Limestone of the Kilbarchan–Howwood area and there is no local representative of the Hurlet Limestone. In the Lora Burn Borehole, the Dockra Limestone is about 15 m thick and consists of medium-grey calcareous mudstone which passes upward into dark grey argillaceous limestone. The fauna is dominated by brachiopods but contains solitary corals in a few places. A thin coal, 0.02 m thick, was present immediately beneath the Dockra Limestone in the borehole. The limestone is exposed in the River Calder [348 589] and has been worked in a small quarry [243 589], 300 m NNW of Garthland.

The strata overlying the Dockra Limestone, consisting mainly of interbedded mudstones and limestones, are equivalent to the Hosie limestones elsewhere in the Greenock district. The sequence of four discrete limestone beds present in other areas cannot be identified at Lochwinnoch. A rooty bed of variable thickness, at the position of the Lillie's Shale Coal, divides the sequence into two units. A marine fauna with brachiopods, crinoids and goniatites is present in the limestones and associated calcareous mudstones. The top of the Lugton Limestone Formation is taken at the top of the uppermost of the Hosie limestones.
                                                         SKM

LIMESTONE COAL GROUP (LSC)

The Limestone Coal Group occurs in three small outcrops on the eastern margin of the district, north of Kilbarchan, and around Lochwinnoch. Nomenclature for the Kilbarchan area follows that of the Central Coalfield, where the group is not divided into formations; the Lochwinnoch area forms part of the Dalry Basin of north Ayrshire, where two forma-

tions are now recognised. Correlation between the two areas is shown in Figure 7.

Kilbarchan area

Only the lowest part of the group, about 20 m thick, is preserved. There is no surface exposure and information comes entirely from boreholes. The strata consist of dark grey bedded mudstones very similar to those in the underlying Lower Limestone Group. They include several clayband ironstones, the thickest of which is the Johnstone Clayband Ironstone. This seam lies about 12 m above the base of the group, is 0.35 to 0.5 m thick, and was extensively mined right up to its outcrop in the Sandholes Basin [40 64] over 100 years ago. The mudstones at the base of the sequence contain a marine fauna including abundant *Posidonia corrugata*. The immediately overlying mudstones contain alternating developments of *Lingula squamiformis* and *Naiadites*, in places in very close association, but the mudstones around and above the Johnstone Clayband Ironstones are usually barren.   IHF

Lochwinnoch area

Kilbirnie Mudstone Formation

The strata overlying the Lugton Limestone Formation consist mainly of mudstones with a number of thin ironstone bands, most notably the Dalry Clayband Ironstone (0.27 m thick), which lies at the horizon of the Johnstone Clayband Ironstone. The Johnstone Shell Bed, a widespread marine horizon, is also present in the Lochwinnoch area and is useful in relating the sequences of the Greenock district with those developed to the south, in Ayrshire.

Dalry Sandstone Formation

The sediments of the Kilbirnie Mudstone Formation are succeeded by a sequence of mudstone, siltstone, sandstone, seatrock and coal, arranged in cycles. These strata are assigned to the Dalry Sandstone Formation. The proportion of sandstone in the cycles increases upwards in the sequence, coal seams becoming more common and strata with a marine fauna becoming rarer. Rocks of the formation are best exposed in the railway cuttings [369 585] on the east side of Castle Semple Loch. Grey mudstone with *Lingula* exposed here is considered to be the local representative of the Black Metals Marine Band, a prominent marker horizon in the Paisley district. Clayband ironstone nodules and beds within the mudstone, referred to in Ayrshire as Logan's Bands, locally reach workable thickness in the Lochwinnoch area.

Towards the top of the Dalry Formation, coals, known locally as the 'Peel Coals' and tentatively correlated with the Main and Smithy coals of north Ayrshire, attain workable thickness and have been mined in places. They are thinner and less persistent than their equivalents around Dalry, in the district to the south.   SKM

CONDITIONS OF DEPOSITION

After a period at the end of Stratheden Group times during which local erosion took place, sedimentation in the district resumed as a result of renewed regional subsidence. The early Carboniferous basin was more extensive than that in which the Upper Devonian was laid down, and in large parts of the southern Midland Valley the basal beds of the Inverclyde Group overlap the entire Stratheden Group to rest upon rocks of Lower Devonian age or older.

The thick, relatively arenaceous Kinnesswood Formation of the Helensburgh–Dumbarton area was probably laid down by braided rivers which flowed into the basin from the north-west. The bulk of the drainage continued eastwards into the Stirling and Fife districts, but part must have flowed southwards in the Firth of Clyde area, depositing the thinner, more argillaceous sequences with southwards-directed cross-bedding at Inverkip and Auchengarth. It may be inferred, therefore, that the basin did not close towards the south as it did during the Upper Devonian period. During the later part of Kinnesswood Formation times, regional subsidence allowed the sea to encroach gradually from the south and east on to the floodplains of the mature river systems. As a result, the Ballagan Formation was laid down in a shallow-water environment with restricted access to the open sea, subject to marked changes of salinity and periodic desiccation. The absence of coarse clastic material indicates that the relief in the source area was subdued.

Just as the onset of marine conditions was gradual, so also was the resumption of fluvial sedimentation towards the end of Ballagan Formation times, probably as a result of rejuvenation of the upland source areas. Deposition of sandstones probably commenced first in the north-western part of the Greenock district but gradually extended southwards and eastwards over the Ballagan Formation mudflats, with the result that the base of the Clyde Sandstone Formation is strongly diachronous.

The abundance of calcrete in the Kinnesswood Formation and lower part of the Clyde Sandstone Formation is consistent with a climate having a mean annual temperature of about 16°C and a moderately seasonal rainfall (cf. Allen, 1974). However, it would appear that, towards the end of Clyde Sandstone Formation times, there was a significant increase in rainfall which promoted the growth of vegetation and raised the height of the water table within the floodplain sediment. As a result of the reducing conditions thereby introduced, the strata of the Overtoun Sandstone Member and Broadlee Glen Sandstone Member are generally grey in colour and contain much finely disseminated carbon as well as plant debris.

Deposition of the Clyde Sandstone Formation in the Greenock district was brought to an end by uplift caused by east–west compressional stresses. North–south-trending folds were formed within the area between the Highland Boundary Fault-system and the Largs Fault-zone. In the core of one of these folds, the Leap Moor Syncline, the thickest Inverclyde Group sequence is preserved. Strata along the crests of anticlinal folds and on the upthrown eastern side of the Largs Fault-zone were subject to erosion and most of the Inverclyde Group was removed in such areas. As a consequence, the volcanic rocks of the Strathclyde Group locally rest upon the Kinnesswood Formation.

After eruption of the lavas and tuffs of the Clyde Plateau Volcanic Formation, the volcaniclastic Kirkwood Formation initially began to accumulate in late Dinantian times on low ground within and adjacent to the thick volcanic pile of the Renfrewshire Hills. As a result of continuing subsidence, however, sediment of nonvolcanic origin began eventually to encroach upon the lava pile. The deposits of the lower part of the Lawmuir Formation consist largely of quartzose sand, but include some volcaniclastic material because the lavas remained uncovered in places till almost the whole of the formation had been laid down. Initially deposition took place mainly in a fluvial environment but later, marine incursions occurred. These attained their maximum extent towards the end of Dinantian times (Lower Limestone Group) and diminished again in the early Namurian (Limestone Coal Group). Between marine episodes, periods when sand was deposited in the channels and on the floodplains of a major river system alternated with periods when the area was colonised by coal-forming vegetation. IBP, IHSH, IHF

SIX

Intrusive igneous rocks

Bodies of intrusive igneous rock in the district range in composition from basalt to rhyolite. The oldest were probably emplaced during the Lower Devonian, about 400 million years ago, into the Dalradian rocks of the south-west Highlands. The youngest were intruded about 52 million years ago, during the Tertiary period. The most extensive and varied suite of intrusions relates to the early Carboniferous eruptive episode which gave rise to the Clyde Plateau Volcanic Formation.

The intrusions vary greatly in size and form. Many consist of near-vertical dykes which in some cases extend across country for many kilometres. Others are roughly cylindrical and probably represent the infillings of volcanic vents. A few of the bodies are sills—gently inclined sheet-like masses approximately concordant with the bedding in the country rock. Although most of the intrusions were emplaced in a molten condition, a number of vent-structures contain fragmental material which includes sedimentary as well as igneous rock types.

LOWER DEVONIAN INTRUSIONS

Two north-east-trending dykes cutting Dalradian rocks have been classed as lamprophyre and are considered to be of Lower Devonian age. Both are sufficiently altered to make classification difficult, but the 4.5 m-wide dyke near Rosneath Church [252 831] contains abundant biotite and may be a kersantite.

LOWER CARBONIFEROUS INTRUSIONS

Large numbers of minor intrusions, both within the Clyde Plateau Volcanic Formation and in the underlying sediments, have strong petrological affinities with the volcanic rocks and are assumed to be contemporaneous with them (Leitch and Scott, 1917; Richey, 1939; Tyrrell, 1917).

Plugs and vents

Several plugs and irregular bodies of vent agglomerate within the outcrop of the Misty Law Trachytic Centre are almost certainly related to this phase of extrusive activity and acted as feeders for the trachytic and rhyolitic lava flows. The most prominent plugs, all of which form steep-sided hills subcircular in plan, are on Knockside Hills, a little to the south of the present district, Little Irish Law [261 594], Black Law [274 594], Box Law [257 611] and Slaty Law [263 611]. The rocks of which they are composed are all porphyritic, with phenocrysts of perthitic sanidine, and less commonly of albite, in a matrix of alkali-feldspar, oxidised mafic minerals and quartz, much of the last being of secon-

dary origin. Almost all of the rocks are brecciated, especially towards the margins of the bodies.

Vent agglomerates are poorly exposed in the Misty Law area but have been recognised around the Black Law plug, on Waterhead Moor and on Hill of Stake. All consist of a mixture of local basaltic and trachytic material and the Hill of Stake vent includes much brecciated glassy basalt. Volcaniclastic rocks in the area around Irish Law and Knockside Hills, formerly mapped as vent agglomerate, are now considered to lie within an apron of extrusive pyroclastic rocks surrounding the vents. True vent agglomerate most probably does occur close to the plugs, but has not been distinguished on the map.

Outwith the centre, plugs are difficult to identify within the mainly basaltic succession of the Clyde Plateau Volcanic Formation, owing to the close correspondence in rock type between lavas and intrusions. Some isolated, upstanding outcrops of fresh basic rock, such as the crags of Jedburgh basalt around Little Craig Minnan [318 644], can be interpreted either as outliers of lava or as plugs. Outcrops more securely identified as plugs occur at Dunconnel Hill [331 595], where the rock is of Markle type, and at Castle Hill [223 639], which is composed of basalt of Dalmeny type. An outcrop of rubbly and highly brecciated Dalmeny basalt at Moniabrock [352 632], measuring 200 m by 100 m, is either a vent agglomerate or a plug.   DS

Four basaltic vents, cutting the Devonian and Carboniferous sediments, beneath the volcanic rocks, have been recognised in the area south of the Clyde. At Underheugh Quarry [203 752], a plug of Markle basalt with well-developed, radially disposed columnar joints (Plate 7) is seen to cut basaltic vent agglomerate with clasts up to 3 m long. On the coast near Auchengarth [193 648] the steep north and south contacts of a large basaltic vent, 1100 m long and 400 m wide, are exposed. The vent agglomerate is made up largely of angular fragments of basalt, but clasts of red and white sandstone and cornstone occur close to the margins of the body. A similar body of basaltic tuff and agglomerate at the Knock [202 628], with a diameter of about 400 m, also contains sandstone blocks and is cut by intrusions of Dunsapie basalt. Small outcrops of basaltic tuff, 1300 m to the south-south-east, are also considered to lie in a vent.   IBP

North of the River Clyde at Dumbarton, a group of vents cutting Lower Carboniferous sediments lies at the south-western end of the Dumbarton–Fintry line of volcanic centres (Whyte and MacDonald, 1974). These include the prominent landmark of Dumbarton Rock [400 745] which consists of a central plug of Jedburgh/Dalmeny basalt (Plate 8) within a screen of vent agglomerate and collapsed sediment (Whyte, 1966; 1980). Farther east, four vents along a line extending north from Dumbuck [422 745] to Ravens Craig [423 758] consist of basaltic vent agglomerate with little sedimentary material, cut by complex multiphase basic

Plate 7 Columnar jointing in volcanic plug of olivine-basalt, Underheugh Quarry [302 752] (D 3512).

intrusions of Jedburgh and Dalmeny type. Small vents at Townend [405 768], Pappert Hill [423 803] and Auchencarroch Hill [423 815] contain both igneous and sedimentary fragments and there are plugs also at Auchencarroch Hill, on Carman Hill [372 795] and on the south-west flank of Ben Bowie.   IHSH

Dykes

Numerous thin, impersistent dykes, from 0.5 to 3 m wide, with compositions corresponding to the full range of rock types encountered in the lava succession, cut the lavas and underlying sediments over the whole area south of the River Clyde. Many of the dykes, especially those seen along the coast between Skelmorlie and Largs, have an approximately east–west or ENE–WSW trend, but dykes with other trends are also common. Dykes are particularly abundant in the lower parts of the lava pile exposed in deep valleys such as Noddsdale and the Greeto and Gogo waters, suggesting that many acted as feeders for fissure eruptions. In particular,

Johnstone (1965) noted that trachytic dykes are abundant within and below the levels of the Misty Law centre, but are virtually absent above. Dykes are notably abundant, and tend to have a predominant NE–SW trend, in areas close to the projected south-westerly continuation of the Dumbarton–Fintry volcanotectonic line (p.19). North of the River Clyde, dykes are abundant close to the vents of the Dumbarton area, in the Murroch Burn and south-west of Renton, where they are intruded into rocks of the Kinnesswood and Ballagan formations. Dominant trends are east–west and NNE–SSW. Farther to the north-west, dolerite dykes with a NE–SW trend occur on Ben Bowie and in Garadhban Forest. A persistent north-east-trending dyke, which crops out on both sides of the Gare Loch at Shandon, has textural similarities to the Lower Carboniferous dykes and has been classed tentatively as a dolerite of transitional Markle/Dunsapie type.

No detailed petrographical study has been carried out on the dykes. The majority are fine grained, nonporphyritic, and therefore difficult to classify in the field. Most dark to

Plate 8 Volcanic plug of olivine-basalt, Dumbarton Rock [400 744] (D 3494).

medium grey varieties have been classed as dolerite or basalt (D), only a few paler varieties being distinguished as mugearite (X^M). Most porphyritic basic dykes are of Markle type (D^M). Paler, greenish grey dykes, which commonly show flow-banding either as colour changes or lines of vesicles, are usually classed as felsic alkaline rocks (O) or, where confirmed by examination of thin sections, trachyte (O^T). However, trachytes can be recognised in the field when they are porphyritic, by their stumpy, rectangular, pinkish phenocrysts of alkali-feldspar, commonly anorthoclase (fO^T). Much paler, pink or cream, highly altered or silicified dykes are classed as felsites (F or, when porphyritic, fF). DS

Larger basic intrusions

An irregular sheet-like body of medium-grained dolerite crops out on Knockencorsan Hill [240 666], where it is seen to cut Inverclyde Group sediments and the Noddsdale Volcaniclastic Beds on opposite sides of a NW–SE fault. Either the intrusion occurs in several leaves or it postdates the fault. It has subsequently been cut by a series of NNE-trending faults related to the Largs Fault-zone. IBP

Within the Misty Law Trachytic Centre, several small bodies of altered doleritic rocks are recognised, the largest of which occur in the Surge Burn [272 608] and on the south-western flanks of High Corby Knowe [271 615]. DS

Larger trachytic and felsitic instrusions

At Craigmuschat [240 775], Gourock, a westward-dipping lenticular body, up to 100 m thick, of heavily altered, haematite-stained feldsparphyric trachyte is intruded into sandstones of the Kinnesswood Formation. At the time of the survey, the rock was well exposed in a large quarry in which the basal contact with bleached white sandstone was seen. The petrography of the intrusion is described by Tyrrell (1917) and by Leitch and Scott (1917), the latter including an analysis. Phenocrysts of albite-oligoclase and perthitic orthoclase are set in a groundmass of orthoclase and quartz, much of the latter being of secondary origin, with iron-titanium oxides, chloritised pseudomorphs after mafic minerals, and apatite. The quarry rock is well known for its geodes containing, among other minerals, crystals of quartz, calcite, baryte and fluorite, the last being rare in central Scotland (Nicholson and Durant, 1986). Feldsparphyric trachytes, similar to the Craigmuschat rock, occur in a group of thin sills between Inverkip and Loch Thom. The most extensive sill is intruded slightly discordantly, close to the boundary between the Stratheden and Inverclyde groups. It is well exposed on the coast at Inverkip War Memorial [200 719] and was 14 m thick in the Everton Borehole. Silicified, feldsparphyric sills which cut trachytic agglomerate at Rye Water Head [255 592] are of similar type and are probably related to the nearby vent intrusions of Knockside Hills and Little Irish Law.   IBP, DS

Microgranodiorite and related rocks of the West Girt Hill – High Corby Knowe area

Much of the high ground to the south of Hill of Stake in the middle of the Misty Law Trachytic Centre is underlain by an equigranular rock of dioritic appearance, which is exposed only sporadically through a thick peat cover over an area 1300 m long and 500 m wide. This was not recognised in previous mapping. Thin sections typically reveal partially albitised laths of oligoclase rimmed by fresh rectangular orthoclase overgrowths, in a groundmass of almost square-sectioned fresh orthoclase, quartz and well-developed micropegmatite. Other minerals include iron-titanium oxides, altered mafic silicate minerals (pyroxene and/or amphibole, plus biotite) and apatite. Variations in grain size occur and in some areas the rock is plagioclasephyric, but in most cases it can be classified as microgranodiorite or quartz-monzodiorite depending on the amount of quartz. The field relations of the mass are difficult to establish owing to the poor exposure and lack of features, but its general form, as inferred from air photographs, suggests a sill-like body some 50 m thick, within, but not necessarily related to, the Misty Law Trachytic Centre.

Coarse-grained altered dioritic rocks, which are relatively abundant as xenoliths in a mugearite exposed in the River Calder [325 620] and penetrated in boreholes, may be derived from a similar but larger intrusion at depth.

LATE CARBONIFEROUS QUARTZ-DOLERITES

A number of east – west or ENE-trending quartz-dolerite dykes in the district are part of a major swarm of late Westphalian to early Stephanian age (295 to 290 Ma) which extends across central Scotland and into the North Sea (Walker, 1935; Macdonald et al., 1981; Russell and Smythe, 1983). The dykes are discontinuous, occurring as segments up to 6 km long with a maximum known width of 30 m, for example in the River Calder at Tandlemuir [335 613]. The disposition of the individual lengths suggests that there are up to five discontinuous dykes in the area south of the River Clyde and four to the north.

Locally the dykes have been emplaced along fault planes, in particular on east – west faults such as the Cloch Fault. They probably postdate major movements on such fractures, although terminations and displacements of the dykes in some areas suggest later fault movements. In the district they are only seen to cut the Clyde Plateau Volcanic Formation and older rocks, but elsewhere in central Scotland the swarm cuts Middle Coal Measures (Westphalian B). In the Nodds-

dale Water, 500 m SSW of Outerwards Reservoir, a 30 m-wide quartz-dolerite dyke is cut by a 24 m-wide dolerite of Tertiary age.

The dykes are usually well jointed, and are composed of fairly fresh blue-grey coarse-grained and equigranular dolerite with diffuse patches of pink quartzofeldspathic material. Most consist essentially of labradorite laths, pseudomorphs possibly after hypersthene, and iron-titanium oxides, all partially enclosed by subophitic augite. The groundmass is usually a mesostasis of quartz and alkali-feldspar, commonly intergrown as micropegmatite. Varieties with a glassy groundmass (tholeiites), which are common in eastern parts of the swarm, have only been recognised in a single dyke length exposed in the Mill Burn [318 658].

EARLY PERMIAN ALKALI OLIVINE-DOLERITES

A sill-complex consisting of several leaves of ophitic olivine-dolerite is intruded into the Lawmuir Formation in the south-east corner of the district. The stratigraphical level at which the sill is emplaced ranges from below the Castlehead Lower Coal to just above the Hollybush Limestone. Numerous exposures occur in the area between Johnstone

Plate 9 Basalt dyke cutting conglomerate of Kelly Burn Sandstone Formation, Wemyss Bay [188 699] (D 1576).

and Kilbarchan, including the large, working Kilbarchan Quarry [407 634] and several small abandoned quarries on the south-eastern limb of the Howwood Syncline. The sill-complex is not known to the north-east of Kilbarchan, where its outcrop is terminated by a major NW–SE fault. The combined thickness of the various leaves within the district is at least 25 m, but aggregate thicknesses of over 50 m are known a short distance to the east.

The sills have been traditionally regarded as members of the Midland Valley teschenitic group on account of the minor amounts of interstitial analcime present in some rocks. However, the amount of analcime is not sufficient to justify this classification and most rocks are best termed ophitic olivine-dolerite or, in a few cases, ophitic analcime-bearing olivine-dolerite. In typical samples, slender laths of plagioclase, iron-titanium oxides and a variable amount (5 to 15 per cent) of euhedral olivine, pseudomorphed in serpentine, bowlingite and chlorite, are enclosed by large ophitic plates of fresh pinkish brown clinopyroxene. Secondary pinkish brown fibrous zeolitic minerals are common.

Several leaves of the complex, cutting deformed strata within the Paisley Ruck, were proved in a borehole north-east of Howwood [405 612] and, in eastern Glasgow, sills of similar composition cut the Middle Coal Measures. Apparent major displacements of the sill-complex occur on east–west and NW–SE faults. Radiometric (K-Ar) mineral dates in the range 273 to 270 Ma from four sills in the Glasgow–Paisley area suggest an early Permian age, contemporaneous with the Mauchline Lavas of central Ayrshire (de Souza, 1979).

TERTIARY DYKE-SWARM

Basaltic dykes of Tertiary age in the district have a dominant NW–SE trend and most are considered to be members of the Mull regional swarm, which extends across southern Scotland and into northern England. The north-eastern edge of the swarm is very sharply defined and within the district is marked by a group of particularly persistent dykes in a zone, 2.5 km wide, extending from Inverkip to Glenlora. To the north-east of this zone, Tertiary dykes are virtually absent, but to the south-west numerous, mainly short lengths occur. Deviations from the dominant north-west or NNW trend are mainly due to local deflections into fault planes, but a few dykes with east–west or ENE trends may belong to a separate subswarm radiating from the central igneous complex of Arran.

The more persistent dykes are commonly 6 to 15 m in width, with well-developed polygonal jointing perpendicular to chilled margins. The smaller dykes are usually 2 to 3 m wide. Most are composed of very fresh hard blue-grey equigranular doleritic rocks which resist erosion. The swarm has not been examined in detail in the district, but the continuation of the swarm to the south has been fully described by MacGregor (in Richey et al., 1930). An olivine-tholeiite dyke, formerly exposed at McInroy's Pier [2185 7680] near Gourock, contains xenoliths and screens of sedimentary rock which have yielded instructive metamorphic assemblages (Herriot, 1971). Some information on the petrochemistry of the swarm is given by Thompson (1982, p.470).

Within the district, most of the dykes of the Mull regional swarm are tholeiitic basalts or dolerites (K^T). They are composed of labradorite laths and augite, with varying amounts of glassy mesostasis which is commonly devitrified and darkened by finely disseminated iron-titanium oxides. Olivine-tholeiites (oK^T) occur rarely. Some of the larger, more persistent dykes are more siliceous quartz-dolerites (K^Q) or andesitic tholeiites (K^A), containing orthopyroxene or pigeonite. Alkali olivine-dolerites (K^C) are less common in the regional swarm and some may belong to the Arran subswarm. Most consist essentially of ophitic purplish titanaugite enclosing labradorite, olivine and iron-titanium oxides. Varieties with small amounts of analcime or zeolites resemble the finer-grained crinanites of Kintyre and Arran.

DS

SEVEN

Structural development

The structural history of the Greenock district can be traced from the early part of the Caledonian Orogeny, in late Precambrian times (c.600 Ma), to the end of the Palaeozoic Era. During much of this time basin development and sedimentation were considerably influenced by periodic movements on fractures within the Highland Boundary Fault-system. This complex structure consists of many elements (Figure 8), some now largely concealed beneath younger rocks but others forming some of the most obvious geological and topographical features of the district. The main NE–SW elements in the system are the Rosneath Fault and the Inchmurrin Fault and its inferred continuation along the Spango valley. These are believed to coalesce to the south-west on the Isle of Bute, forming a single structure which probably links with the Corloch Fault on Arran. The Largs Fault-zone, which is considered to be a major splay of the system, trends generally SSW from Greenock via Nodds-dale to Largs as a series of NNE–SSW faults. One of these continues SSW to Farland Head and may link with other major fractures, such as the NE–SW Cumbrae Fault and the north-east–south-west Dusk Water Fault, beneath the Firth of Clyde.

EARLY CALEDONIAN EARTH MOVEMENTS

During the early part of the Caledonian Orogeny, in what has been termed the Grampian event, the Dalradian rocks were deformed into a series of major folds (F1), cleaved and subjected to regional metamorphism (p.5). Recent radiometric dating of the Ben Vuirich Granite of Perthshire (Rogers et al., 1989) implies that the initial deformation and nappe formation occurred prior to 590 Ma. If so, the Grampian event must be divided into two main phases—an early phase at c.600 to 595 Ma involving nappe formation, generation of cleavages (Sl and S2) and metamorphism to greenschist grade; and a later phase at around 520 to 490 Ma when further deformation occurred and peak metamorphic conditions were attained. The earlier phase is responsible for the creation of the Tay Nappe and Aberfoyle Anticline, but there is little sign in the Greenock district of the later Grampian event. During a later phase of the Caledonian Orogeny (470 to 440 Ma), block uplift of the Dalradian terrain led to the formation of the 'downbend' (p.5), which was followed closely by compression and development of the S4 cleavage.

A series of NNE-trending faults, which displace the rocks in the axial zone of the Aberfoyle Anticline, appear to com-

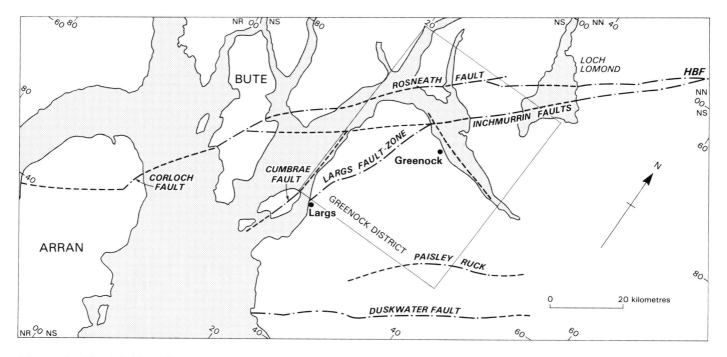

Figure 8 The Highland Boundary Fault-system and other major fractures in the Firth of Clyde area. (HBF = Highland Boundary Fault-system)

bine sinistral strike-slip movements with vertical displacements. The age of these brittle structures, which may relate to movements on the Highland Boundary Fault-system, is not known, although they can be no older than the formation of the downbend. The date of initiation of the important Highland Boundary Fault-system is unknown but it has been suggested that it already existed as a basement fault during deposition of the Dalradian.

The varied assemblage of rocks known as the Highland Boundary Complex, which includes strata of Middle Ordovician age (Curry et al., 1984), was juxtaposed against the Dalradian in late Ordovician or earliest Silurian times. There is no clear consensus regarding the process by which this was achieved but obduction of the strata during the closure of the Iapetus Ocean has been suggested. The present relations of the Highland Boundary Complex with the Dalradian and Devonian rocks would appear to have resulted from movement on fractures in the Highland Boundary Fault-system over a considerable period, perhaps commencing at the time of the downbend.

LATE CALEDONIAN AND EARLY HERCYNIAN EARTH MOVEMENTS

There is no evidence in the district which relates to sedimentary or structural events during the Silurian period but late-Silurian sediments a little to the south at Farland Head [178 484] are generally fine grained and show no sign of having been deposited in the vicinity of a highland front. During the Lower Devonian, however, the Greenock district was part of a major north-east-trending subsiding basin centred on the Midland Valley and flanked to north and south by upland areas. It has been suggested, largely on hypothetical grounds (Bluck, 1978; 1980), that formation of this basin involved strike-slip movements on the Highland Boundary Fault-system. Palaeomagnetic measurements indicate, however, that the Highlands and the Midland Valley have occupied similar positions in relation to each other since Ordovician times (Harte et al., 1983).

It is likely that the input of coarse detritus into the Midland Valley basin during the Lower Devonian was maintained by rejuvenation of the Highland source area by intermittent uplift on the Highland Boundary Fault-system. Although Lower Devonian strata almost certainly encroached upon the Highlands, it is probable that in the Greenock district they do not now extend beyond the line of the South Inchmurrin Fault and its inferred south-westwards extension (Figure 8; cf. George, 1960, fig. 8).

During the Middle Devonian, compressive earth movements brought about marked changes in the palaeogeography of central Scotland. The Lower Devonian rocks were folded into north-east-trending structures and there were major thrust movements on elements of the Highland Boundary Fault-system. An especially large displacement on the South Inchmurrin Fault caused the erosion of any Lower Devonian rocks that may have been deposited to the north. The Lower Devonian within the basin was deeply eroded, particularly along anticlinal axes, and continuing uplift prevented the deposition of any Middle Devonian sediments. The direction of the palaeoslope within the

Midland Valley was reversed with the result that the predominantly south-west-flowing Lower Devonian rivers were replaced during the Upper Devonian by an axial drainage directed towards the east-north-east.

During most of the Upper Devonian period, tectonic activity was subdued. Deposition of the Stratheden Group took place under conditions of gentle regional subsidence, with intermittent uplift of the Highlands on fractures within the Highland Boundary Fault-system. Towards the close of the period, the Midland Valley sedimentary basin had reached maturity and, in the Greenock district, encroached extensively upon the Highlands. At the end of Upper Devonian times, however, there was a renewal of tectonism, with reactivation of elements of the Highland Boundary Fault-system, the main displacement being on the Rosneath Fault. Uplift to the west, similar to that which affected the Corloch Fault on north Arran (George, 1960), caused the removal by erosion of any Upper Devonian strata. Within the basin, there was upthrow to the west on the Largs Fault-zone, an important north-north-east-trending belt of fractures which was to play a major role in Carboniferous times. The uppermost portion of the Stratheden Group in the western part of the district was removed by erosion and the earliest Inverclyde Group strata locally succeed unconformably.

Deposition of the Inverclyde Group in the Midland Valley in early Carboniferous times took place under quiet tectonic conditions. Slow regional subsidence continued and the floodplain deposits of the Kinnesswood Formation were progressively covered by coastal sabkha sediments of the Ballagan Formation. Later rejuvenation of Highland source areas caused fluvial sediments to advance southwards and eastwards over the coastal mudflats. Towards the end of Inverclyde Group times, the sedimentary sequence within the Greenock district began to be affected by east–west compressive movements. The Largs Fault-zone was reactivated but its throw was reversed, the area to the east being raised by several hundred metres and subjected to erosion. The strata lying between the Largs Fault-zone and the Rosneath Fault were folded into the north–south-trending Leap Moor Syncline and adjacent anticlines (Figure 1), the crests of which were later eroded.

Within the district, the lavas and associated volcaniclastic sediments at the base of the Strathclyde Group rest upon an eroded surface of rocks which range from high in the Clyde Sandstone Formation to the Kinnesswood Formation. This widespread early Carboniferous volcanicity may relate to a tensional phase initiated as a reaction to the preceding compression. To the east, in the Glasgow district, most of the Clyde Plateau Volcanic Formation can be related to north-east-trending lines such as the Dumbarton–Fintry line (Hall and Forsyth, in preparation). The differences between the lava sequences in the Renfrewshire Hills, the Beith–Barrhead area and the Kilpatrick Hills may be due in part to differential movements of blocks bounded by major crustal fractures, such as the Clyde Fault and the Paisley Ruck. The effects of such block faulting upon the build-up of sedimentary sequences in the Midland Valley during the Carboniferous have long been known.

MAIN HERCYNIAN EARTH MOVEMENTS

Earth movements during the later Carboniferous gave rise to faults, most of which fall within one of four principal trends: north–south, east–west, NE–SW and NW–SE. The numerous faults which cut the volcanic rocks of the Renfrewshire Hills belong mainly to this period. The Largs Fault-zone also was activated again, this time with an easterly downthrow of several hundred metres. The North Inchmurrin and Rosneath faults were active, rocks of Upper Devonian and early Carboniferous age being let down between them in a graben-like structure. Among other important NE–SW structures of later Carboniferous age are the Camis Eskan Graben, north of the Clyde, the faulted syncline centred on the Spango valley and the complex Paisley Ruck. Some details of the last-named structure are given by Hinxman et al. (1920, p.45). The emplacement of Devonian strata within horst-like structures to the south-west of Ben Bowie and in Greenock [290 755] may have taken place at about the same time.

It is probable that most of the east–west faults, some of which are intruded by quartz-dolerite dykes, were in being by the end of the Carboniferous period. Several faults of this trend, such as the Muirshiel and Cloch faults, have throws of several hundred metres and have a marked effect on outcrop distribution. At least some of the east–west faults may have been active during Permian or later times as they apparently displace the olivine-dolerite sills of the Kilbarchan area which are considered to be of early Permian age.

The latest of the faults are those of NW–SE trend, many of which have large throws and can be traced for considerable distances. Members of the group, such as the Kilbarchan Fault, cause major displacements of the upper boundary of the Clyde Plateau Volcanic Formation and of the early Permian olivine-dolerite sill-complex. The faults of this group belong to a major fracture system which was probably initiated in the late Carboniferous but which continued to be active during Permian and later times. They may have controlled some of the late Palaeozoic magmatism in the Midland Valley and undoubtedly controlled the siting of important mineral veins, as at Muirshiel, which are possibly of Mesozoic age.   IBP

EIGHT

Quaternary

In common with the rest of Scotland, the Greenock district was covered by ice-sheets on a number of occasions during the Quaternary Era as the climatic regime in northern Europe alternated between the arctic and the temperate. The effects of glacial activity upon the landscape are everywhere visible, the most striking features being the drowned ice-scoured valleys of Loch Long, the Gare Loch and the Firth of Clyde.

Many of the superficial deposits in the district reveal by their characteristic form or internal structure that they were laid down by glacier ice or its meltwaters. The most prominent of the landforms, such as the deep sea-lochs, are probably the product of repeated glaciations over a long period. Most of the deposits, on the other hand, relate to the most recent glacial episodes. In the period since the last glaciers disappeared, about 10 000 years ago, the Greenock landscape has changed relatively little.

OUTLINE OF LATE QUATERNARY HISTORY

The Greenock district was last entirely occupied by ice in late Devensian times, when an ice-sheet, which reached its farthest limits at about 18 000 years Before Present (BP), covered Scotland and northern England. The subsequent retreat of the ice-sheet initially took place in very cold climatic conditions. This is indicated by assemblages of molluscs, foraminifera and ostracods, similar to communities presently inhabiting the seas around north Norway and Spitsbergen, which have been recovered from the Errol Beds, a sequence of glaciomarine sediments deposited at this time around the Scottish coasts (Peacock, 1975). The ice-recession during the period of arctic climate is considered to reflect diminished precipitation. The rate of ice-retreat may have increased after about 13 500 BP, when a marked climatic amelioration initiated the Windermere Interstadial (Table 8), a period during which the climate was relatively mild although still colder than at present. Faunas recovered from the marine sediments of this period are indicative of boreal (that is, cold temperate) conditions.

At the start of the climatic amelioration, the margin of the late Devensian ice-sheet still lay south and west of the present district. This is indicated by the absence of the Errol Beds in the Firth of Clyde north of Largs, where the oldest marine deposits contain a fauna consistent with the warmer climate of the Windermere Interstadial. As a consequence of the northward retreat of the ice-front during the Interstadial, the sea eventually gained access to the lower Clyde valley and the Glasgow area, and by about 12 500 BP (conventionally, before 1950) the Greenock district was probably free of ice.

Because the land was deeply depressed by the weight of ice, the relative sea-level in newly deglaciated areas initially was far higher than at present, as shown by occurrences of beach shingle and other marine deposits at heights of 30 m or more above OD around the Clyde estuary. As a consequence of isostatic rebound, however, the sea subsequently fell for a time, reaching levels considerably below OD.

During the Loch Lomond Stadial, which began about 11 000 BP (Table 8), very cold climatic conditions returned to Scotland and glaciers advanced into the Greenock district from a centre in the western Highlands. Positions maintained by ice-fronts in the Gare Loch, in Glen Fruin and around Loch Lomond are marked by ridges of moraine (Simpson, 1933). During the Loch Lomond Stadial the sea-level rose, as water bound up in continental ice-sheets returned to the oceans and, towards the end of the period, attained a position higher than at present. The final retreat of the glaciers from the district was rapid, following a marked climatic improvement that commenced at about 10 300 BP.

During the Flandrian, alluvial deposits were laid down by streams and rivers, and were terraced as the drainage adjusted to changes of sea-level. These consisted of an initial fall from the position reached towards the end of the Loch Lomond Stadial, followed by a marked eustatic transgression, in the course of which the sea rose to an even greater height, probably by about 6500 BP, as in the Flanders Moss area (Sissons and Brooks, 1971), before falling by stages to its present level. As a result of the initial regression, peat developed in low-lying coastal areas and continued to accumulate, with interruptions, until modern times. In the last few thousand years, peat has also grown widely in poorly drained upland areas as mainly thin but extensive deposits blanketing the high ground and, more thickly, in basins. Most recently, human activity has brought about considerable changes to the landscape, especially along the Clyde estuary, where extensive areas of made ground have been produced.

The chronology of the events outlined above (Table 8) depends to a large extent on radiocarbon datings carried out on shells, peat, wood and other organic materials. The accuracy of the dates thus obtained has been questioned (e.g. Sissons, 1981; Sutherland, 1979; Lowe and Gray, 1979), but radiometric dating is commonly the only method whereby an absolute chronology can be achieved. By carrying out determinations on a series of samples through a sequence, on materials carefully selected to reduce the possibility of contamination, results are obtained which appear to provide a reasonable indication of the general age of the deposit. Dates on single specimens, however, must be treated with reserve, and radiocarbon dates cannot be expected to define short-lived events.

MAIN LATE DEVENSIAN ICE-SHEET

Erosional features

The principal valleys in the district, many of which have a general north–south or NW–SE trend, were doubtless

Table 8 Relations of the late Devensian and early Flandrian marine and associated deposits

		Time scale (years BP)	Movements of relative sea-level	Principal events	Stratigraphical divisions showing presumed time-ranges		Climatic conditions
FLANDRIAN		6000	Overall fall	Main Postglacial Shoreline + 15 m OD	Deposits of Main Postglacial Shoreline		Climatic Optimum
		? 6500	Rise				
		8000					Boreal
			Fall	Peat growth across exposed tidal flats	Peat of Linwood area		
		10 000					High Boreal
LATE DEVENSIAN	Loch Lomond Stadial	10 320	Rise	High Buried Shoreline >5 m OD	Balloch Formation	Clyde Beds	Arctic
				Advance of glaciers			
		10 920		Depositional of marine silts and clays with *Portlandia arctica*			
	Windermere Interstadial		Fall	Erosion of deep channels	?hiatus		?High Boreal
			Approximately stable	Main Lateglacial Shoreline and platform (−2 to −5 m OD)			Mid- to low-Boreal
		12 000	Fall from marine limit	Deposition of marine silts, clays and sands	Linwood Formation		High Boreal
				Deposition of laminated marine silts and clays	Paisley Formation		
		13 000		Onset of deglaciation			

scoured and deepened by ice on more than one occasion during the Quaternary Era. Such valleys characteristically have U-shaped cross-sections and are straight, with truncated spurs and hanging tributary valleys. Typical examples are the Firth of Clyde and the valleys occupied by Loch Long and the Gare Loch, all of which have been locally excavated to depths in excess of 50 m below OD (Deegan et al., 1973). A drift-filled channel, which reaches a depth of 62 m below OD in the Inverleven Borehole (Appendix 3), extends southwards from Loch Lomond beneath the alluvium of the Leven valley before turning south-eastwards and passing along the north side of Dumbarton Rock to join the Clyde estuary. Other major valleys affected by the passage of ice include Glen Fruin, the valleys of the Spango and Greeto waters, Strathgryfe and Noddsdale (Plate 10).

Most, if not all, the major valleys lie along important fault lines in the solid rocks. The planes of weakness offered by the numerous small faults which cut the lavas of the Renfrewshire Hills have also been exploited by glacial scour, as is well seen in the tract of ground between Loch Thom and the Port Glasgow – Kilmacolm Road. Ridges with an east-south-east trend in the area east of Helensburgh are probably mainly carved from solid rock although some may consist of till. Dumbarton Rock is a crag-and-tail feature, the 'tail' of which descends eastwards beneath marine and alluvial deposits. Smaller-scale erosional features, almost certainly to be attributed to the last glaciation of the area, are the numerous polished and striated rock surfaces, which are particularly common in the volcanic terrain south of the River Clyde.

From the orientation of glacial features, four trends of ice-flow can be identified in the Greenock district as a whole, although one, or at most two, usually predominates in any given area (Figures 9 and 10). Thus, in the largely drift-free high ground south of the River Clyde (Figure 9, Area A), east – west and north – south trends are indicated by the evidence mainly of striae. In the low ground flanking the Clyde estuary (Figure 9, Area B), the glacial features, comprising both striae and drumlins, are mainly aligned in a direction 20 – 30 degrees south of east. In the valley at Lochwinnoch, in the south-east part of the district (Figure 9, Area C), the chief trend is NE – SW. There is little direct

Plate 10   Trap features in Clyde Plateau Volcanic Formation, modified by glacial action. Brisbane Glen [203 617] (D 3672).

evidence within the district from which the sense of the ice-flow can be deduced, but *roches moutonées* at a number of localities suggest that, in general, the ice flowed from the west and north. This is consistent with the eastward carry of erratic blocks in ground east of the Greenock district, from, for example, the Lennoxtown essexite (Peach, 1909), and the widespread distribution in central Aryshire of shelly till which includes materials originating in the Firth of Clyde (Eyles et al., 1949).

It has been suggested, for example by Rose (*in* Jardine, 1980), that the various orientations of glacial features arise as a result of deflection of the ice-streams by major landforms, such as the prominent escarpment along the south side the Clyde estuary. However, the principal flow-directions inferred for the Greenock district can be recognised on a regional scale. The east–west, north–south and NE–SW trends, for instance, can all be identified as far south as central Ayrshire (Richey et al., 1930; Eyles et al., 1949). Also, whereas many of the drumlins in the Glasgow district are elongated in an ESE–WNW direction, some striae on the high ground of the Kilpatrick and Campsie hills have a north–south to NNW–SSE trend. The extent to which the flow directions in the Greenock district were deviated by the topography is probably minor, because the orientation of striae at numerous localities, as around Strathgryfe No. 1 Reservoir [280 718], indicates that the ice commonly was constrained to flow up steep hillsides. This in turn suggests that ice-thicknesses at the time of formation of these erosional features was very great.

The implied scale of the ice-streams suggests that they were related to stable configurations of the ice-sheet as it developed and decayed. Features with NE–SW, east–west and NW–SE trends are all represented along the Clyde coast, and it may be inferred that the position of the ice-shed (the central axis of the ice-sheet) lay some distance farther west. This is likely to have been while the late-Devensian ice-sheet was close to its maximum extension. Evidence of the relative ages of these ice-flow directions is limited, but Eyles et al. (1949) considered that striae trending NW–SE postdated others which were aligned in a NE–SW direction and, according to Richey et al. (1930), glacial features with NE–SW and NW–SE trends in the Irvine district are later

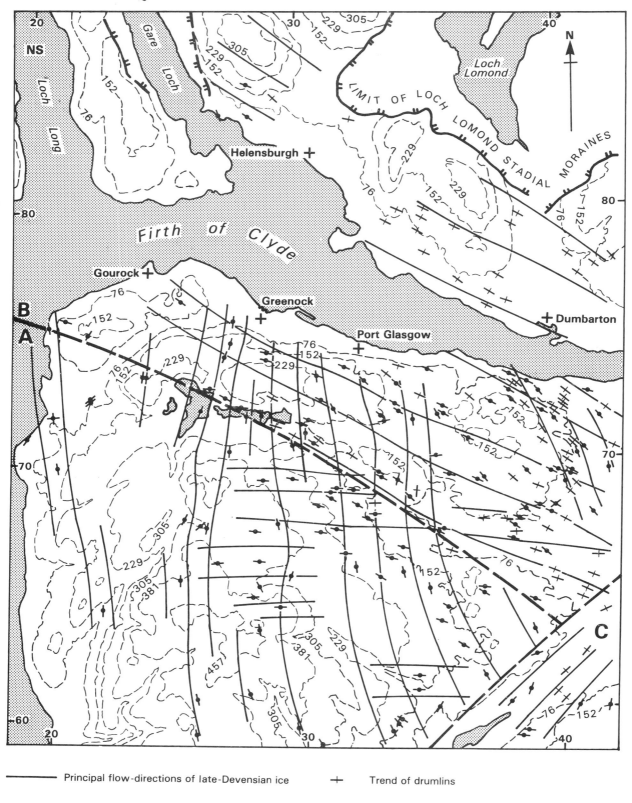

Figure 9 Principal flow-directions of late Devensian ice in the Greenock district, as indicated by trends of glacial striae and drumlins. Areas bounded by thick lines and designated A, B and C correspond with those on Figure 1. Contour heights are given in metres.

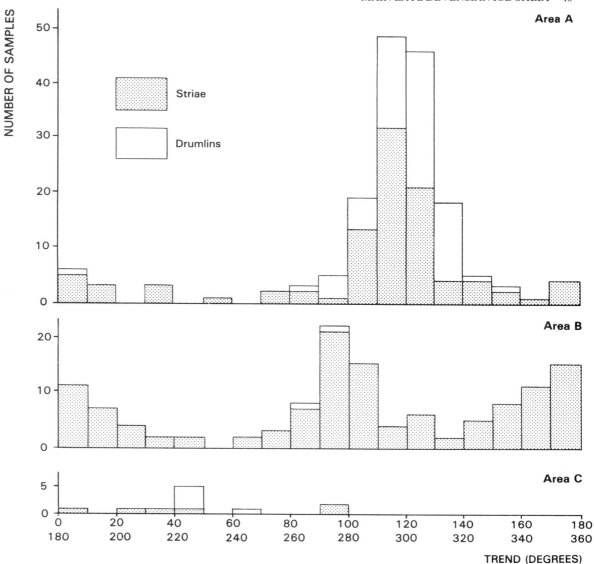

Figure 10 Trends of glacial features in areas A, B and C delineated on Figure 9.

than features with an east–west trend. Taken as a whole, the evidence suggests that the west–east set of striae is the oldest and that it and the other sets with an easterly component were cut while the axis of a major ice-sheet, perhaps formed by the coalescence of ice-caps centred on the western Highlands, the Southern Uplands and Northern Ireland, lay to the west of the Clyde coast. The striae of north–south trend, which on Duchal Moor [291 673] appear to be superimposed upon an east–west set, may have formed last, as Highland ice encroached farther into the Midland Valley, following decline of the southern ice-centres.

Glacial deposits

The typical deposit laid down directly by the ice-sheet consists of till—a compact diamicton composed of angular rock clasts of local and distant origin set in a sandy, silty clay matrix. On the higher ground the till cover is thin and patchy but reaches to the top of the highest hills. In the lower ground, the till deposit is more continuous except in a narrow strip along the coast where it has usually been removed

by marine erosion. However, a small thickness of grey clayey till survives in places in the cliff behind the present beach near Cloch Point [2056 7606], and was visible, at the time of the earlier surveys, resting upon a striated rock surface [202 254], 450 m south of Cloch Lighthouse. At Wester Ardoch Farm [358 764], east of Cardross, a thin deposit of red-brown till rests, at a height of about 10 m above OD, upon a rock platform which is backed by a cliff 10 m high (Browne and McMillan, 1984).

In the Helensburgh–Dumbarton area and along the coast south of the Firth of Clyde, the till is usually clayey, grey-brown in colour, and contains clasts mainly of Dalradian rock types. Farther east, the till is brown or red-brown in colour and contains angular blocks of Upper Devonian sandstones and conglomerates, and lavas and sandstones of Carboniferous age. Even in the lower ground, the thickness of the deposit rarely exceeds 5 m except in valleys such as Shielhill Glen and Noddsdale, or in drumlins—streamlined mounds elongated in the direction of ice-flow. Drumlins are best-developed along the north side of the Clyde estuary between Helensburgh and the Vale of Leven, where they have a general east-south-east trend (Figure 9). South of the Clyde drumlins are less common but there are good examples in the area between Barochan Cross [404 692] and the Clyde coast at West Ferry. It is notable that some of the drumlins have similar trends to, and are presumably contemporaneous with, striae considered to have formed while the ice-sheet was close to its maximum thickness.

In the Erskine Borehole (Appendix 3), there is an exceptional occurrence of two layers of till separated by overconsolidated red-brown laminated clay. The position of the lower till, beneath the laminated clay, suggests that it is pre late Devensian or even pre-Devensian in age, although there is a possibility that the laminated clay has been rafted into its present position.   IBP

The only fauna possibly associated with the main late Devensian ice-sheet was in the Balloch Borehole, where a microfauna dominated by the cold water foraminifera *Elphidium clavatum* and *E. albiumbilicatum* was recovered from till between depths of 51.00 m and 52.60 m (Appendix 1). In a sparse poorly preserved macrofauna, the presence of the bivalve *Chlamys islandica* is evidence that the marine sediments now incorporated within till were laid down in a harsh and inhospitable climate, perhaps during the advance of the main late Devensian ice-sheet.   DKG, IBP

Meltwater drainage channels and deposits

Channels cut by meltwaters flowing at the base of an ice-sheet, or by water spilling from an ice-impounded lake, are present in the Greenock district but are not common. Channels obliquely descending the till-covered hillslopes in the area north of Largs were probably cut by meltwaters flowing beneath the ice near the eastern margin of the receding Firth of Clyde glacier. The valley through Lochwinnoch may for a time have acted as a spillway for water impounded in the upper Clyde valley although more generally the water drained eastwards towards the North Sea. A deep spillway [321 835] carried the overflow into the Clyde from a lake which developed when Glen Fruin was dammed by an ice-lobe during the Loch Lomond Stadial. Most of the existing stream courses and river valleys in the district probably also functioned as meltwater channels.

Fluvioglacial deposits of sand and gravel associated with the main Devensian ice-sheet are also uncommon in the district. Examples include small ridges of sand and gravel (eskers) containing pebbles mainly of Carboniferous lava a little south of Kilmacolm [359 679] and north of Bridge of Weir [394 668]. The latter deposit has been almost removed by quarrying. Both eskers appear to have been laid down by meltwaters escaping to the south and east, presumably flowing into the upper Clyde valley. Sand and gravel deposits [210 600] at Largs are probably fluvioglacial outwash terraces, the lower parts of which have been reworked by the late Devensian sea.

The origin of the small area [914 698] of sand and gravel north of Wemyss Bay is problematical. The deposit, which is truncated at its upper and lower ends, falls southwards from about 32 m above OD to about 26 m above OD. It thus descends to a level considerably below that at which the sea stood when it flooded the Glasgow area, following melting of ice occupying the Clyde estuary. In consequence, the Wemyss Bay spread can have a fluvioglacial origin only, it would appear, if a considerable mass of dead ice remained in the valley of the Spango Water and around Inverkip after the ice in the Firth of Clyde and the Clyde estuary had been breached by the calving of bergs in the deeper waters along the axial channel.

Associated marine deposits

Marine deposits of late Devensian age in the Firth of Clyde area have long been included in the general category of Clyde Beds (Table 8), which embrace sediments laid down under varying climatic conditions over a considerable time span. The lower parts of the Clyde Beds are characterised by a fauna of molluscs, foraminifera and ostracods indicative of sea temperatures considerably colder than today but significantly less frigid than those prevailing during deposition of the Errol Beds of eastern Scottish coastal regions (Peacock, 1975; Paterson et al., 1981). The mid- to high-boreal climatic conditions inferred from the fauna are believed to have become established following the important climatic amelioration at about 13 500 BP (Ruddiman and McIntyre, 1973), which marks the start of the Windermere Interstadial. This interpretation is consistent with the majority of the radiocarbon dates (13 150 BP to about 12 500 BP) obtained from molluscan shells in the lower part of the Clyde Beds sequence, although a somewhat older radiocarbon date (13 780 BP) on shells from a Clyde Beds sequence at Hawthornhill [375 760] has recently been reported (Browne et al., 1983). The early part of the Clyde Beds was laid down in the temperate conditions appropriate for the Windermere Interstadial, but the evidence of marine faunas and radiocarbon dates at a number of sites (Peacock et al., 1978; Browne and Graham, 1981) indicates that the later part was deposited under distinctly colder conditions, in the period between 11 000 BP and 10 000 BP, that is, during the Loch Lomond Stadial.

It has recently been proposed (Browne and McMillan, 1985) that the younger, colder part of the Clyde Beds should be assigned to the Balloch Formation and that the older, less frigid part be subdivided into a lower Paisley Formation, consisting of colour-laminated clays and silts, and an upper Linwood Formation, composed mainly of grey silts and clays. The last two units both contain the characteristic mid- to high-boreal fauna but in the lower division it is generally rather sparse. A thin layer of shelly bouldery gravel intervening between the Linwood and Balloch formations has been termed the Inverleven Formation. The Paisley and Linwood formations are described here but the Balloch and Inverleven formations are described with the deposits of the Loch Lomond Stadial (p.46).

Paisley Formation

A deposit of closely laminated clay, silt and sand in various shades of grey, brown and red, which lies at the base of the Clyde Beds at many localities, is assigned to the Paisley Formation. The deposit is rarely more than one metre thick but reaches a thickness of 2 to 3 m at Ardyne (Peacock et al., 1978) and of 3.3 m in the Linwood Borehole (Appendix 3). The lower part of the Paisley Formation tends to be coarser than the upper part and contains laminae with small angular rock fragments. Larger rock clasts, up to 0.1 m long, scattered through the deposit, and occurring also in the overlying Linwood Formation, were probably dropped in from floating ice-masses. Rapid deposition by bottom traction currents and from turbid meltwater plumes in the neighbourhood of a retreating ice-front has been suggested (Peacock, 1981). As the age of the Paisley Formation is related to the age of local deglaciation, its relationship with the succeeding Linwood Formation is probably diachronous.

From published descriptions, it would appear that the laminated clays and silts of the Paisley Formation were encountered in excavations at Greenock for the Garvel Park Dock [295 758] (Robertson, 1883) and at the nearby New Graving Dock [307 752] (Bishop and Dickson, 1977). They were also found at the site of Inverkip Power Station (R G Ward, personal communication) and at levels up to 29.5 m above OD in excavations [375 760] at Hawthornhill, Dumbarton (Browne et al., 1983). Currently, the Paisley Formation is exposed only in the Geilston Burn, Dumbarton [341 777], where it is about 0.6 m thick (Rose, *in* Jardine, 1980, pp.26–27).

From the evidence at Hawthornhill, it is apparent that when deposition of the Paisley Formation commenced, in the period immediately after deglaciation, the sea stood at least 30 m higher than its present level. Water depths along the axis of the Clyde estuary and the Leven valley must accordingly have exceeded 100 m at this time but diminished later as the sea-level fell from the marine limit.   IBP

Sparse marine faunas of very restricted diversity were recovered from the Paisley Formation in the Inverleven and Linwood boreholes (Appendix 1). In the former the macrofauna included a fragment tentatively identified as belonging to the cold water bivalve *Chlamys islandica*. Some echinoid and barnacle fragments were also present. In the Linwood Borehole, samples of the Paisley Formation contained no macrofossils or ostracods. In a very restricted assemblage of foraminifera, relatively common *E. clavatum* was accompanied by sparse *Cassidulina reniforme*, *Pyrgo williamsoni* and *Quinqueloculina seminulum*. On the whole the faunas point to a cold and unfavourable marine environment.

At Hawthornhill, Dumbarton [375 760] a thin deposit of marine clay contains a shallow marine molluscan assemblage in which the molluscs *Mytilus edulis* and *Littorina saxatilis* are accompanied by cold water indicators *Macoma calcarea* and *Nuculana pernula*, proving that temperatures were lower than at present, although probably mid- to high-boreal rather than arctic. This deposit is immediately overlain, not necessarily conformably, by a freshwater deposit which, although considered to be Flandrian, contains in addition to the gastropods *Gyraulus laevis* and *Lymnaea peregra* the cold water bivalve *Pisidium obtusale lapponicum*.   DKG

Linwood Formation

The Linwood Formation consists of clayey silt and sand, grey and grey-brown in colour with black sulphide-rich patches. There is a small proportion of angular rock debris, including clasts up to 0.15 m long, which is believed to have originated as dropstones. Some of these have been colonised by barnacles and serpulids. In general, the deposit is poorly laminated, perhaps because of bioturbation. The formation

is the most fossiliferous part of the Clyde Beds and has yielded the bulk of the fossils ascribed to that division.

In the Linwood Borehole (Appendix 3), the clayey silts and sands of the Linwood Formation were about 23 m thick. A radiocarbon date of 11 800 BP (Peacock, 1971) suggests that at least the lower part of a 66 m thick sequence of grey silts and clays penetrated by a borehole at Dumbarton Distillery is referable to the Linwood Formation. At Geilston Burn (Rose, *in* Jardine, 1980, p.27), a sequence up to 4 m thick of fossiliferous grey silts and clays resting upon the Paisley Formation is here referred to the Linwood Formation. Numerous boreholes drilled along the waterfront in Greenock and offshore have also proved considerable thicknesses of grey silts and clays of the Clyde Beds but in the absence of faunas or radiocarbon dates, the proportion that should be allocated to the Linwood Formation is not known. From the radiocarbon dates yielded by shells incorporated in morainic deposits of the Loch Lomond Stadial south of Loch Lomond and at Rhu [265 840] (Jardine, 1980), it may be inferred that the Linwood Formation was also laid down in Loch Lomond and the Gare Loch.   IBP

Faunas of the Linwood Formation have been recovered at the type locality, at Langbank [331 785], from several localities in the Dumbarton area, and in blocks of sediment incorporated in till deposits in a stream section at Townhead Farm, Helensburgh [316 828] (Browne, McMillan and Hall, 1984). Extensive excavations last century at Garvel Park Dock [295 758] provided the best exposures of the Linwood Formation in the area and faunas from these have been discussed in a number of papers. Unfortunately, some collections from the site (Crosskey and Robertson, 1871, pp.32–35) appear to have combined Flandrian fossils with material from the Linwood Formation. Also, the discovery of a fully arctic fossil assemblage (Scott and Steel, 1882, pp.274–283) indicates the presence at the site of deposits laid down during the Loch Lomond Stadial (Peacock, 1981, pp.222–236; 1987, pp.93–103). On the whole, the Linwood Formation is much better known in the areas immediately west and east of the present study area where there have been a number of rewarding excavations. One of the largest was the oil platform construction site at Ardyne (Peacock et al., 1978, pp.1–25; Graham and Wilkinson, 1978, pp.1–17).

The faunas tend to be significantly more diverse than in the Paisley Formation, with individual species occurring in greater abundance (Appendix 1). Conditions were clearly more favourable than during deposition of the Paisley Formation, but whilst the faunas contain a number of species which would be found in Scottish waters today they also include a proportion of taxa which have a more northerly distribution at present, showing that temperatures remained significantly lower than they are now. Typical cold water molluscs of the Linwood Formation are the bivalves *Chlamys islandica*, *Macoma calcarea*, *Nuculana pernula*, and *Yoldiella fraterna*. Microfossils whose presence affords evidence of low water temperatures include foraminifera, of which *E. clavatum* is usually numerically dominant, and ostracods, of which *Cytheropteron latissimum* is one of the most frequently occurring. The tendency overall is for faunas to be more diverse in the outer Clyde estuary than upstream, presumably reflecting variations in salinity levels in the contemporaneous estuary.   DKG

When deposition of the Linwood Formation commenced, in the period after about 12 800 BP, the sea was still considerably higher than at present but was falling as a result of isostatic uplift. The bulk of the formation may have been laid down in the form of a marine delta advancing seawards in the Clyde estuary as the sea fell from the marine limit.

LOCH LOMOND STADIAL

During the early part of the Loch Lomond Stadial, at about 11 000 BP, the climate became colder and glaciers nourished by an ice-cap in the south-west Highlands entered the northern part of the Greenock district by way of the valleys now occupied by Loch Lomond, the Gare Loch and, presumably, Loch Long. The Loch Lomond glacier also entered lower Glen Fruin where it impounded a lake about 6 km long, with a surface level at about 114 m above OD (Price, 1983). Meltwaters spilling from the lake cut a deep channel at Helensburgh as they flowed into the Firth of Clyde. The outer limits of conspicuous morainic landforms south and west of Loch Lomond and on the hillslopes flanking the Gare Loch mark positions where the glacier fronts were stabilised at or near the farthest limit of this glaciation. However, the local occurrence of blocks of shelly Clyde Beds in the till outside the moraine limits indicates that the glaciers may have reached beyond the moraine line during the period of maximum extension.

In addition to widespread deposits of till, the Loch Lomond Stadial glaciers laid down sand and gravel as englacial deposits, in the form of mounds (kames) and esker ridges, and as proglacial outwash spreads. The latter are best developed in the Vale of Leven south of Loch Lomond, where they rest upon a thick series of marine silts and clays, the Balloch Formation, which is considered to represent the early, distal outwash from the Loch Lomond glacier.

The position of the sea-level during the Loch Lomond Stadial has been much debated. Evidence from boreholes at Balloch and Inverleven in the Leven valley (Browne and Graham, 1981) indicates that the sea may have been at a level considerably below OD at the beginning of the period but rose during deposition of the Balloch Formation to a position above present sea-level, perhaps by about 10 300 BP. Marked overconsolidation of the upper part of the formation in the Balloch Borehole (Browne and Graham, 1981) indicates that the final advance of the Loch Lomond glacier, presumably to the line of the arcuate moraine ridges, took place after that date. The recession of the glaciers of the Loch Lomond Stadial and the general disappearance of glacier ice from Scotland apparently took place rapidly as a result of a marked climatic amelioration which commenced at about 10 300 BP. The evidence of pollen assemblages at Dubh Lochan on east Lomondside (Stewart et al., 1984) implies that retreat of the Loch Lomond glacier from the district had occurred by early in the *Juniperus-Empetrum* pollen zone, the beginning of which has been dated at about 10 000 BP.

Glacial deposits

The chief deposit laid down by the Loch Lomond Stadial glaciers consists of a somewhat loose, brown, sandy till with angular rock fragments of local origin and an admixture of far-travelled clasts. The deposit commonly contains lenses and patches of sand and gravel, particularly near its outer limits, as in the narrow, clearly defined ridges along the north side of Glen Fruin and ascending the valley sides above the Gare Loch, north of Rhu (Rose, *in* Jardine, 1980). South of Loch Lomond, the till locally contains shells derived from deposits of Linwood Formation traversed by the glacier during its passage down the loch.

Unlike the till of the main Devensian ice-sheet, which gives rise only to smooth sheets and drumlins of considerable size, the Loch Lomond Stadial deposits are commonly formed into the irregular mounds and elongate ridges described by the term 'moundy moraine'. The outer limits of these deposits are visible as arcuate ridges along the southern and western flanks of Loch Lomond, in Glen Fruin and on the hillslopes above the Gare Loch. These ridges mark a former position of the glacier fronts, but not necessarily their maximum extent. Thus, the discovery in till near Helensburgh on the Gare Loch (Browne et al., 1983), of blocks of Linwood Formation containing shells, including *Arctica islandica*, which have yielded radiocarbon dates of 11 700 BP and 11 000 BP, suggests that the maximum position reached by the Gare Loch glacier was some distance beyond the ice-limit marked by the moraine ridges at Rhu.

The extensive deposit of sand and gravel in the Leven valley south of Loch Lomond was partly laid down in association with stagnant ice of the retreating Loch Lomond glacier, as shown by the occurrence of kames, esker-ridges and kettleholes. Ice-contact deposits in this area are surrounded and partly buried by a spread of outwash sand and gravel, the surface of which has been terraced in response to later changes of sea-level. As shown by the Balloch Borehole, the terraced outwash rests upon the marine silts and clays of the Balloch Formation.

Outwash deposits related to a glacier of this period are present also in the Gare Loch at Rhu (Rose, *in* Jardine, 1980, pp.31–37), where glacially deformed marine silts and clays are overlain by up to 4 m of fluvioglacial sand and gravel which is capped by Flandrian beach deposits [265 840]. The fluvioglacial deposit consists of a lower steeply cross-bedded unit which is overlain by flat-lying beds, the junction varying in height between 2.3 m above OD and about 5 m above OD. The cross-set deposit consists mainly of sand and gravel but includes beds of sand and silt. Both fluvioglacial units are locally folded and sheared and are believed to have been overridden by the advancing Gare Loch glacier.

Marine deposits—Balloch and Inverleven formations

In a few places within the Greenock district, marine deposits are known which are considered on the evidence of their faunas and radiocarbon dates to have been laid down during the Loch Lomond Stadial (Browne and Graham, 1981). The deposits were proved by recent BGS boreholes at Balloch and Inverleven (Appendix 3). The former penetrated 39.75 m of brown clay and silt to which the name Balloch Formation has been given (Browne and McMillan, 1985). This rested, at a depth of 39 m below OD, upon a thin layer of shelly bouldery gravel termed the Inverleven Formation. The site of the Balloch Borehole lies within the Loch Lomond arcuate moraine and the upper part of the silt and clay deposit is heavily overconsolidated as a result of glacial loading. The sequence proved farther downstream in the Leven valley by the Inverleven Borehole (Appendix 3) is generally similar. However, at this site, which lies outside the limit reached by the Loch Lomond glacier, the sediments are normally consolidated.   IBP

Fossils from the Loch Lomond Stadial were identified in the Balloch and Inverleven boreholes (Appendix 1). Both contained very restricted molluscan faunas in which the presence of the arctic bivalve *Portlandia arctica* indicates that water temperatures were appreciably lower than in the Paisley or Linwood formations. In both boreholes the micro-faunas are characterised by moderately diverse foraminiferal assemblages, but the ostracods are very restricted and of distinctly cold water aspect.   DKG

Shell material from the Inverleven Formation has yielded radiocarbon dates of 11 960 BP and 10 350 BP at Inverleven and 10 920 BP at Balloch. The oldest of these suggests the possibility that the gravel layer is, in part, a lag deposit produced as tidal currents reworked the Linwood Formation which formerly occupied the buried channel of the Leven valley. It is possible, however, as Browne and Graham (1981) have suggested, that the gravel is a condensed deposit which represents the whole of the Linwood Formation. Current action would have been most vigorous while Loch Lomond was empty of ice and available as a tidal reservoir, but it is unlikely that tidal scour alone could have maintained a channel to a depth of 50 m below OD and a period of lowered sea-level may be indicated. Advance of a glacier into the loch would have progressively reduced the tidal scour in the Leven valley and eventually allowed deposition of the overlying clay and silt of the Balloch Formation seaward of the glacier. The height of the uppermost part of the Balloch Formation suggests that by the end of the Loch Lomond Stadial the sea-level was at least as high as at present.

Evidence obtained last century from excavations at Garvel Park, Greenock, and recently reassessed by Peacock (1981), is compatible with the late Devensian sea-level movements inferred from the boreholes in the Leven valley. Thus, in the western part of the Garvel Park Dock, grey silts and clays contain a boreal fauna (Robertson, 1883) of Windermere Interstadial type and are probably Linwood Formation. The top of the deposit was at about 2 m below OD. At the eastern end of the dock, grey silts and clays with a gravelly layer at the base reached a depth of about 11 m below OD; the fauna, which includes *P. arctica*, indicates that the deposits are probably referable to the Balloch Formation. There is no information regarding the nature of the contact between the two sets of deposits, but it is possible that the presumed Balloch Formation occupies a channel eroded in the older late Devensian deposit.

LATE DEVENSIAN SEA-LEVELS

Changes of sea-level in relation to the land during the decay of an ice-sheet are the result of the interaction between the eustatic movements of sea-level, as water is added to or subtracted from the oceans, and the isostatic recovery of the land from the depression caused by the former ice-load. At localities close to the centres of large ice-sheets, the extent of the glacio-isostatic depression considerably exceeds the amount by which the sea has been eustatically lowered. In consequence, when the sea eventually gains access to such areas, its level is initially high but usually falls rapidly because the rate of isostatic uplift, generated by the earlier thinning of the ice-sheet by downwasting, already exceeds the average rate of eustatic sea-level rise. In such circumstances, the initial level of the sea following deglaciation equates to the marine limit, that is, the highest postglaciation sea-level. Even in such central areas, the rate of isostatic recovery may nevertheless be matched at times by exceptionally rapid eustatic rise during periods of accelerated ice-melt worldwide. Shoreline features, later raised and tilted by the continuing differential isostatic uplift, may be formed at such times while the sea is maintained at a generally stable level.

In the Greenock district, evidence relating to the late Devensian sea-level movements is limited and the account that follows draws upon information from adjoining areas in the Midland Valley of Scotland. Within the area adjacent to the Firth of Clyde, evidence on the position of the marine limit has been obtained at a few localities only. Near Toward Point [110 690], a little outwith the Greenock district, the highest late Devensian raised beach is at a height of about 38 m above OD (Peacock et al., 1978). Sutherland (1984) considered the marine limit to be at about 41 m above OD in lower Loch Long but provided no evidence. According to Rose (*in* Jardine, 1980), marine deposits and features reach heights of 28 m above OD at Dumbarton, 26 m above OD at Geilston and about 24 m above OD at Helensburgh. However, from occurrences of shelly marine deposits up to a height of 29.5 m above OD at Hawthornhill, Dumbarton [375 760], Browne et al. (1983) deduced that the marine limit in that area may have been as high as 35 m above OD. This is similar to the height reached by deposits of fine-grained well-rounded gravel, possibly beach shingle, on the slopes within and adjacent to Erskine Golf Course [435 725], a short distance east of the district. It has been suggested (Peacock, 1971; Rose, 1975) that the marine limit in Glasgow is above 30 m above OD and may even have been as high as 40 m above OD (Browne et al., 1983).

Deglaciation of the Firth of Clyde area occurred after the climatic amelioration of 13 500 BP. The marine limit may therefore lie at a level corresponding to that of the prominent feature known as the Main Perth Shoreline (Sissons and Smith, 1965), which is considered to have formed at that time in the Tay and Forth valleys (Paterson, 1974; Armstrong et al., 1985). The shoreline has not, however, been identified around the Firth of Clyde. In the Forth and Tay valleys, formation of the Main Perth Shoreline was followed by a considerable drop of sea-level. A comparable fall of sea-level from the marine limit during the period following deglaciation has been postulated for west coast areas on the basis of water depths deduced from the evidence of molluscan faunas found at Ardyne and elsewhere (Peacock et al., 1977; Peacock et al., 1978). The hillslopes adjacent to the Firth of Clyde are mainly steep and rocky and only in the area between Helensburgh and Dumbarton are there raised shoreline features (Rose, *in* Jardine, 1980) which may date from this early post-deglaciation fall of the sea from the marine limit. No data regarding the altitude or gradient of these features have been published.

According to Peacock et al. (1977; 1978), the post-deglaciation fall of the sea in western Scotland was halted at about 12 000 BP and, in the period that followed, its level was stabilised at a position somewhat higher than at present. At Ardyne, a bench was eroded at a height of about 5 m below OD. The corresponding sea-level position in the Tay and Forth valleys is probably marked, near OD, by the Main Lateglacial Shoreline and associated erosion surface. These prominent features were considered by Sissons (1969; 1974) to have formed under a cold climate during the Loch Lomond Stadial. However, Browne, Graham and Gregory (1984) has suggested that, as they must have formed under conditions of relatively stable sea-level, they more probably date from a period of ameliorated climate, from 12 000 to 11 000 BP, during the later part of the Windermere Interstadial, when accelerated eustatic rise of sea-level may have counterbalanced the effect of continuing isostatic uplift.

On the evidence of deep channels eroded into late Devensian marine deposits in the Tay and Forth valleys, it has been argued (Paterson et al., 1981; Browne, Graham and Gregory, 1984) that during the Loch Lomond Stadial the sea was at a level considerably lower than at present. Erosion of a deep channel in response to a lowering of sea-level at, or just before, the start of the Stadial would explain the evidence from the Vale of Leven where the Inverleven Formation, with a radiocarbon date of 10 920 BP, occurs to a depth of more than 50 m below OD in the Inverleven Borehole (Appendix 3), whereas the older Linwood Formation is apparently present at a much shallower depth in the nearby borehole at Dumbarton Distillery. The evidence at Garvel Park Dock (p.47) also is consistent with a phase of channel erosion. The sea is considered to have risen during the Stadial, while the Balloch Formation was being laid down, to culminate at a position above OD before the end of the period. The equivalent position in the Forth valley is marked by the High Buried Shoreline, for which a date of about 10 300 BP has been proposed (Sissons, 1966). During the last part of the late Devensian period, isostatic uplift brought about a major fall of sea-level. This continued into early Flandrian times.

There is evidence from south-western Norway, an area with a high isostatic uplift rate, of a marine transgression that commenced during the Allerød Interstadial, a period of relatively mild climate between 12 000 and 11 000 BP, and culminated at about 10 400 to 10 300 BP (Anundsen, 1978; 1985; Krzywinski and Stabell, 1984). This evidence is consistent with the inference that a major rise of relative sea-level took place during the Loch Lomond Stadial in the estuaries of central Scotland. It has been suggested that the rise of the sea to the High Buried Shoreline was caused by redepression of the crust beneath the load of Loch Lomond Readvance ice (Sissons, 1969, p.39), but this appears unlikely if the transgression was actually initiated at or before the start of the Stadial. It is possible, therefore, that the transgression may be the consequence of a marked rise of eustatic sea-level. As this implies a major input of meltwater into the oceans from ice-sheets in other parts of the globe, it is of interest that there is now a body of evidence (e.g. Davis and Jacobson Jr, 1985) to indicate an accelerated retreat of the North American ice-front during the period from about 11 000 to 10 000 BP.

MAIN ROCK PLATFORM

By far the most prominent marine feature in the district is a broad, rockcut bench with a steep, commonly cliffed back feature at a height ranging from 7 m above OD in the south-west (Gray, 1978) to about 12 m above OD in the north-east. The feature is present along the coast everywhere, except on the steep sides of the Gare Loch and Loch Long, and has been identified in Loch Lomond (Rose, in Jardine, 1980). It is particularly well developed along the coast from Largs to Inverkip, where the cliff backing the platform in many places reaches heights of more than 30 m above OD and has caves cut in it, as at Skelmorlie.

The rock platform is usually covered by a thin veneer of beach sand and shingle of Flandrian age and in the past has been dated to this period. The platform was, however, correlated with the Main Lateglacial Shoreline and associated erosion platform of the Forth valley by Sissons (1974), who considered that it was formed by wave and frost action during the Loch Lomond Stadial. However, a pre late Devensian age for the rock platform by Browne and McMillan (1984), on the grounds that at Wester Ardoch [358 764], which lies outwith the known limit of the Loch Lomond Readvance, the feature is overlain by till. Several of a series of boreholes drilled recently during site investigations for a proposed outfall sewer east of Gourock Bay [254 775] also proved red-brown till resting upon a rock platform and overlain by Flandrian raised beach deposits.

A late Devensian age for the rock platform cannot be reconciled with the sea-level movements proposed for the period.

FLANDRIAN

The landscape of the Greenock district has been little modified during the 10 000 years since the start of the Flandrian Stage. The deposits laid down during this period include spreads of alluvium on the floodplains of streams and rivers, peat in basins and as thin but extensive spreads on hillslopes, and raised beach sand and shingle. In the area flanking the Clyde estuary, quarries, rock cuttings and tunnels, and deposits of made ground, bear witness to human activities.

Alluvium

Alluvium in the form of floodplain deposits and terraced spreads is found along many of the streams and rivers, the most extensive being in the Leven valley, in the valleys of the Spango Water, the River Gryfe, the Noddsdale Water and the Black Cart. In many parts of the district, but especially in the area between Port Glasgow and Bridge of Weir, isolated patches of alluvium occupying hollows mark the sites of former lakes, now silted up. In such places, the deposits are generally of silt and mud, commonly peaty. Along the rivers, the alluvium is more variable in character, consisting of gravel, sand and silt in varying proportions, with local peat.

There is little detailed information about the nature of the alluvial deposits. Boreholes drilled for site investigations in the valley of the Spango Water, south-west of Greenock, show up to 16 m of alluvium, consisting mainly of silt and sand with some gravel. Locally the sequence contains brown peat, which can be several metres thick.

Marine deposits

From evidence outwith the area, the sea is known to have fallen during the early part of the Flandrian (e.g. Sissons and Brooks, 1971). In the Forth and Tay valleys, peat, locally termed the Sub-Carse Peat, developed upon abandoned marine flats as the sea fell to a level close to OD during the period from about 9000 to 8000 BP, and was in large part drowned as a result of a major eustatic transgression which reached its maximum level, at the Main Postglacial Shoreline, by about 6500 BP (Sissons and Brooks, 1971).   IBP

In the Greenock Outfall Borehole No. 20 [2624 7786] the fauna appears to be predominantly Flandrian (although many of the taxa occur in both Flandrian and Clyde Beds sediments) but the presence of the cold water bivalve *Yoldiella fraterna* confirms that at least some of the material is derived. The fauna from the Gourock Pier Boreholes [244 777], on the other hand, is a diverse estuarine assemblage, of undoubted Flandrian age throughout. There are no cold water indicators and it contains molluscs such as the gastropod *Bittium reticulatum* which is never found in very cold water. The microfaunas from these localities have not been identified.

DKG

Within the Greenock district the Flandrian transgression is considered to have reached a height of 14 m above OD at Rhu and about 15 m above OD at Loch Lomond (Rose, *in* Jardine, 1980). Thin deposits of sand and shingle were laid down on the Main Rock Platform or upon its till cover but, because the supply of sediment was generally limited, the deposit was built up to the contemporaneous sea-level in only a few places. Sand and gravel of the Main Postglacial Shoreline deposit is present to a height of about 8 m above OD at Rhu and more than 10 m above OD at Geilston (Rose, *in* Jardine, 1980).

Peat

During the period of marine regression in the early part of the Flandrian, peat developed on low ground around Linwood, a little east of the district. The base of the peat ranges in height from 12 m above OD at Clippens [433 654], where a radiocarbon date of 9231 BP was obtained, to about 4.5 m above OD (Bishop and Coope, 1977). Peat growth continued in the area, possibly with interruptions as a result of marine transgression, until after 3500 BP. The earliest part of the deposit is equivalent to the Sub-Carse Peat of the Forth and Tay valleys.

Much of the ground above a height of about 300 m above OD is covered by a deposit of brown fibrous peat. Usually this is less than one metre thick but locally reaches thicknesses of two or more metres, as at Blood Moss [215 695], on Duchal Moor [275 670] and in the area [265 605] between Slaty Law and Irish Law. Peat contained in small basins in the lower ground, as in the neighbourhood of Dowries [272 706], in places exceeds 1.3 m in thickness.

Made ground

Extensive deposits of made ground along the south shore of the Clyde estuary between Greenock and Port Glasgow were laid down mainly during the expansion of the shipbuilding industry in the nineteenth century. There are also deposits of made ground in the Leven valley, laid down during development of industrial sites in and around Dumbarton. These deposits are usually only a few metres thick, consisting of rubble and materials made available by excavations. Made ground deposits at factory sites in the valley of the Spango Water are up to 5 m thick.

A number of old quarries on both sides of the river have been used for the disposal of domestic waste, the principal sites north of the Clyde being at Renton [368 790], Fairy Knowe [312 838] and Bonhill [399 790], and, on the south side of the river, at Knocknairshill [297 746] and West Kilbride [337 723].   IBP

NINE

Economic geology

Exploitation of the mineral resources in the district was formerly much more extensive than at present and bulk materials such as crushed rock, sand and gravel, brickclay, building stone, limestone and slate were worked in numerous small pits and quarries. Copper minerals were mined at sites near Greenock and Lochwinnoch and considerable quantities of baryte were extracted at Muirshiel. Currently, only Kilbarchan Quarry [408 636] and Dumbuckhill Quarry [420 747] are working on a full-time basis, producing crushed rock aggregate, although some rock is extracted at Underheugh Quarry [203 751] from time to time.

IBP

METALLIFEROUS MINERALS

Baryte

Muirshiel Mine [282 649] was one of three baryte mines in southern Scotland (Muirshiel, Gasswater and Glen Sannox), all now closed, which collectively accounted for about one third of the UK annual production between 1946 and 1966. The mine was active intermittently from the 1830s to 1969. Total production was about 300 000 tonnes, mostly from post-1942 activity, with a maximum output of 17 000 tonnes per annum in the early 1960s. A full description of the mined deposit and other baryte occurrences in the Renfrewshire Hills is given by Stephenson and Coats (1983) who expand earlier accounts by Wilson et al. (1922), MacGregor (1944) and Hobson (1959). More detailed mineralogical and geochemical studies are described by Moore (1979a; 1979b). The mine is situated on the faulted northern margin of the Misty Law Trachytic Centre, where two veins of high purity (more than 95 per cent) banded pink and white baryte were worked, both in massive open-jointed trachyte flows. The worked veins trend NNE–SSW and east–west, and were up to 6.5 m wide.

A trial mine in the nearby Berryglen Burn investigated a NW–SE vein, 1.2 m wide. Elsewhere more than forty baryte veins have been recorded and almost all occur in the vicinity of the Tertiary regional dyke-swarm. They are particularly abundant close to the north-eastern edge of the swarm, where the dykes are most numerous. The majority of veins occur within the Misty Law Trachytic Centre where they occupy fractures having a wide variety of directions. Outside the trachytic centre, in the overlying basaltic sequence, baryte veins and irregular pods are closely associated with major ESE-trending faults or with east–west- or ENE-trending late Carboniferous quartz-dolerite dykes. Such veins are generally less pure than those in the trachytes and commonly contain minor sulphide minerals. Only one trial working is known, from near Heathfield Farm [325 621].

Copper

Copper mineralisation occurs in a wide variety of environments in the Renfrewshire Hills and several small copper mines and prospects were worked in the 19th century (Wilson, 1921; Stephenson and Coats, 1983). Copper occurred as disseminations of malachite in the Clyde Sandstone Formation at two localities south of Gourock, Drumshantie [240 766] and Larkfield [249 762]. Opencast pits were in operation before 1810 and small scale underground workings were carried out at Drumshantie from 1873 to 1875. The malachite occurred as segregations and thin veins, characteristically associated with carbonised plant debris, and a little bornite was recorded. Similar minor disseminations are recorded in shales near the Cloch Lighthouse [202 755] and in sandstones and cementstones at several localities east of the Leven valley.

Two mines at Kaim [345 614], near Lochwinnoch, were worked intermittently between about 1848 and 1877 with moderate success (800 tons of ore in 1861). The copper occurred with baryte, quartz and calcite in several veins up to 3 m wide extending for 800 m along the southern margin of an east-north-east-trending quartz-dolerite dyke that cuts mugearite and Markle lavas. The main copper-bearing minerals were malachite with bornite, chalcopyrite and possibly some chalcocite. Other small copper trials were made in the banks of the River Calder, mostly prior to 1869. Adits were recorded at Calderbank Bleachfield [351 598], Sandy Linn, [346 603], Reikan Linn [335 613] and a post-1869 adit is still visible near Bridgend [348 595]. Most of these probably followed thin malachite-bearing baryte veins, some of which have an east–west trend, in mugearite lavas.

Age of the mineralisation

Hydrothermal activity was obviously prevalent during eruption of the Clyde Plateau Volcanic Formation and may even have occurred earlier, as indicated by the presence of copper in sediments near Gourock. However, many veins follow ENE- to ESE-trending faults and some permeate late Carboniferous quartz-dolerite dykes, as at Kaim. These baryte and copper veins probably relate to an episode of late Carboniferous or early Permian hydrothermal activity which was responsible for the emplacement of many veins of similar type in central and southern Scotland.

Most of the larger baryte veins in the district, particularly those within the Misty Law Trachytic Centre, occur in fractures trending between SSE and south-east. Almost all the veins occur within the area occupied by the Tertiary dyke-swarm, many members of which have trends within this range, and it has been suggested, therefore, that the baryte veins also are of Tertiary age (for example, MacGregor, 1944). However, there is little evidence in north-west Europe of hydrothermal activity during this period, and it is possible that the mineral veins and the dykes were injected at different times into a system of fractures which may date from late Carboniferous times. Radiometric (K-Ar) determinations from vein gouge clays at Muirshiel, ranging in age from 240 to 213 ± 3 Ma (Moore, 1979b), suggest that the main phase of the mineralisation may have occurred during the Triassic period and was approximately contemporaneous with the emplacement of baryte at Strontian, in Glen Sannox and in northern England. DS

BULK MINERALS

Crushed rock aggregate

The potential resources of crushed rock aggregate in the area are restricted to the metamorphic rocks in the Rosneath peninsula and north-west of Helensburgh, the lavas of the Clyde Plateau Volcanic Formation in the Renfrewshire Hills, and dolerite and trachyte intrusions (see Merritt and Elliot, 1984).

The metamorphic rocks north-west of a line from Kilcreggan to Rossdhu House [361 896] on Loch Lomond consist in the main of hard, cleaved sandstones interbedded with slates and thin impure limestones. Because of the cleavage, the rock tends to break into flaky fragments, giving a very poor shape to the aggregate and its strength is not uniform. The rock can, however, be used for fill in foundations or as large blocks in seawalls. The quarry at Camsail [258 823], Rosneath, supplied stone for the Rhu to Faslane road improvement and the upgraded section of the road at Clynder. IHSH

The lavas of the Clyde Plateau Volcanic Formation, in the area to the south of the Clyde, are mainly of basic composition and are included in the 'Basalt' Group of the British Standard Group Classification for roadstone. Most of the lavas have been altered to soft, rather rotten rocks, which lack strength and contain soft clayey minerals undesirable in most aggregates. Lava has, however, been used for fill and as road bottoming and there are large, disused quarries in the neighbourhood of Port Glasgow, for example at Knocknairshill [297 742]. A large quarry [418 694] near Bishopton provided material for the approach roads of the Erskine Bridge.

Intrusions of igneous rock may provide better materials for aggregate, because the rock tends to be uniform and fairly fresh. Two main categories of intrusive rocks have been or are being exploited in the area, namely dolerite, which is included in the 'Basalt' Group of the British Standard Classification for roadstone, and trachyte, which is included in the 'Porphyry' Group. Dolerite occurs as dykes, sills and, in a few cases, as plugs; trachyte usually occurs as sills. IBP

On the north side of the Firth of Clyde, a plug and associated vent agglomerate and being worked at Dunbuckhill [420 747]. A few dykes and small sills of dolerite, some of which were worked in a small way in the past. On the south side of the river, an east–west quartz-dolerite dyke was formerly worked near Kilmacolm [365 692] and Tertiary dolerite was quarried north of Largs [228 649]. The

dykes range in thickness from about 15 to 20 m and are unlikely to afford sufficient reserves for a modern quarry. IHSH

An olivine-dolerite sill, some 15–20 m thick, intruded in Lower Carboniferous sediments, is currently being worked in a large quarry [409 635] near Kilbarchan. Here, too, reserves are limited. The large, disused Craigmuschat Quarry [240 775] in an irregular sill-like body of feldsparphyric trachyte is now used for waste disposal. Underheugh quarry [203 750] near Cloch Point is currently working a small dolerite plug but reserves are limited.

Limestone

In the district, limestone occurs in the Lower Limestone Group and the Lawmuir Formation, in the Ballagan Formation, and in the Kinnesswood Formation. There are also thin beds of limestone in the Dunoon Phyllite.

The Hurlet Limestone, a shaly limestone up to 12.7 m thick at the base of the Lower Limestone Group, was worked extensively near Bridge of Weir [399 659 and 399 654] and in the Howwood area. Although it is much thinner, the Hollybush Limestone in the Lawmuir Formation was worked in quarries [412 666] in the same area.

Limestone in the Ballagan Formation was worked for local agricultural purposes near Camis Eskan [335 821]. Limestone in the Kinnesswood Formation has been worked near Carman Reservoir [381 787], at Inverkip [205 718] and in Shielhill Glen [240 720]. There were also small workings in thin limestone bands within the Dunoon Phyllite in Glen Fruin.

None of these operations was of more than transient local significance, and none of the occurrences of limestone within the district have any potential as limestone resources. IBP, IHSH

Sand and gravel

The principal resources of sand and gravel lie to the north of the River Clyde in moundy and terraced deposits laid down in association with the Loch Lomond glacier. The sand and gravel terraces lying to the west of Loch Lomond have been extensively worked around Midross [359 859]. Terraced outwash has also been extracted in the Leven valley at Balloch [385 823], and at Dalmonach [396 804] raised beach sand and shingle was worked. The extensive sand and gravel deposit in lower Glen Fruin has also been exploited, for example at Daligan [327 844] and at Fingals Hill [336 847] south of Callendoun, but considerable resources remain. A preliminary assessment of the sand and gravel deposits in the vicinity of Loch Lomond has recently been carried out as part of a wider study (Aitken and Armstrong, 1985). IHSH

South of the Clyde, sand and gravel was worked in a small esker ridge [396 668], near Houston.

Brickmaking materials

The clays and silts of the Linwood Formation in the area to the east of the district have been extensively used in the manufacture of bricks and tiles. Within the district, the Linwood Formation and the lithologically similar Balloch Formation are mainly confined within the buried channels of the Clyde and Leven.

It is possible that mudstones in the Ballagan Formation could be used for brickmaking (Elliot, 1985). The rock contains about 8 per cent of dolomitic limestone as hard seams and layers of nodules, and a small amount of gypsum. Both materials would have to be eliminated prior to brick manufacture. Similar mudstones, but without the dolomitic limestone bands, occur in the Lower Limestone Group and in the Lawmuir Formation, near Bridge of Weir. None of these raw materials has been used for brickmaking in the past in this area and their worth in this respect is unknown.

IBP

Slate

Slate was worked at one time in two small quarries on the hillside west of Clynder. The slate waste from one quarry is used from time to time for making up forestry roads. The slate is rather inconsistent in quality and cannot be considered as a resource.

Sandstone

Within the district there are several sandstone beds which have been used for building stone. None of the original quarries is still in use but many of them could be reopened for small quantities of stone. Bright red or carmine sandstone from the Stratheden Group near Renton [386 791], Bonhill [398 787] and Dumbarton [388 760] was used locally for houses and churches. The Kinnesswood Formation provided a pale red to grey freestone near Helensburgh [311 840, 313 830] and at Dumbarton [391 753]. White sandstone of the Clyde Sandstone Formation was worked for local housing in Gourock, Greenock and Port Glasgow. There are several disused quarries along the cliff forming the inland margin of the raised beach platform west of Gourock [222 768 to 235 773]. There were also quarries on Gourock Golf Course [225 763], in Shielhill Glen [235 722], at Everton [219 761] and near Port Glasgow [398 747].

IBP, IHSH

Coal

The Hurlet Coal at the top of the Lawmuir Formation was worked near Bridge of Weir and at Howwood, and the alum shale in the roof of the coal was probably worked at the same time. The quantity of coal remaining in the district is insufficient to be economically worked and the alum shale is too thin a deposit to interest modern industry. The Sandholes Coal at Sandholes [414 644] and the Quarrelton Thick Coal near Kilbarchan were also exploited to a small extent. The Lillie's Shale Coal of the Lower Limestone Group, which contains a seam of oil shale as well as coal, was probably mined more than 100 years ago in the area between Houston and Kilbarchan. The Peel Coals, of Limestone Coal Group age, have been worked near Lochwinnoch.

Ironstone

The Johnstone Clayband Ironstone was extensively mined over a century ago in the Sandholes basin [40 60] but the clayband ironstones of the Lower Limestone Group do not appear to have been exploited. There is some evidence, however, that the Logan's Bands clayband ironstones, in the Limestone Coal Group, were worked in the Lochwinnoch area.

IHF, SKM

Peat

The extensive deposits of peat south of the River Clyde are mostly thin but may reach considerable thicknesses in basins such as at Blood Moss [215 695] and on Duchal Moor. In the past, the peat has been exploited on a small scale for domestic purposes.

Groundwater

Current exploitation of the groundwater resources of the area is limited to a few private wells, used mainly for domestic and agricultural purposes. The best prospects for groundwater are considered to be the sandstones of the Stratheden and Inverclyde groups, from which yields of up to 10 litres per second might be expected.

IBP

REFERENCES

AITKEN, A M, and ARMSTRONG, M. 1985. The sand and gravel deposits of the Lomondside–Balfron area, Strathclyde and Central Regions. *Open-File Report, Inst. Geol. Sci., Edinburgh.*

— and SHAW, A J. 1983. A preliminary study of the sand and gravel deposits of Lomondside. *Open-File Report, Inst. Geol. Sci., Edinburgh.*

ALLEN, J R L. 1974. Sedimentology of the Old Red Sandstone (Siluro–Devonian) in the Clee Hill area, Shropshire, England. *Sediment. Geol.*, Vol. 12, 71–167.

ANUNDSEN, K. 1985. Changes in shore-level and ice-front positions in late Weichsel and Holocene, southern Norway. *Norsk. Geogr. Tidsskr.*, Vol. 39, 205–225.

ARMSTRONG, M, and PATERSON, I B. 1970. The Lower Old Red Sandstone of the Strathmore Region. *Rep. Inst. Geol. Sci.*, No. 70/12.

— PATERSON, I B, and BROWNE, M A E. 1975. Late-glacial ice limits and raised shorelines in east-central Scotland. In *Quaternary studies in north-east Scotland*. GEMMELL, A M D (editor). (Department of Geography, University of Aberdeen.)

— PATERSON, I B, and BROWNE, M A E. 1985. Geology of the Perth and Dundee district. *Mem. Br. Geol. Surv.*, Sheets 48W, 48E, 49 (Scotland). 108 pp.

ASPEN, P. 1974. Fish and trace fossils from the Upper Old Red Sandstone of Dunbartonshire. *Proc. Geol. Soc. Glasgow*, Vol. 113, 4–7.

BISHOP, W W, and COOPE, G R. 1977. Stratigraphical and faunal evidence for Lateglacial and early Flandrian environments in south-west Scotland. In *Studies in the Scottish lateglacial environment*. GRAY, J M, AND LOWE, J J (editors). (Oxford and New York: Pergamon Press.)

— and DICKSON, J H. 1970. Radiocarbon dates related to the Scottish late-glacial sea in the Firth of Clyde. *Nature, London*, Vol. 227, 480–482.

BLISS, G M, and GRANT, P R. 1979. Possible stratiform stromatolites in the Dunoon Phyllites. *Scott. J. Geol.*, Vol. 15, 63–64.

BLUCK, B J. 1978. Sedimentation in an orogenic basin: the Old Red Sandstone in the Midland Valley of Scotland. *In* Crustal evolution in northwestern Britain and adjacent regions. *Geol. J. Spec. Issue* No. 10

— 1980. Evolution of a strike-slip controlled basin, Upper Old Red Sandstone, Scotland. *In* Sedimentation in oblique-slip mobile zones. *Spec. Publ. Int. Assoc. Sediment.*, No. 4.

BRADBURY, H J, HARRIS, A L, and SMITH, R A. 1979. Geometry and emplacement of nappes in the Central Scottish Highlands. In *The Caledonides of the British Isles—reviewed*. HARRIS, A L, HOLLAND, C H and LEAKE, B E (editors). *Spec. Publ. Geol. Soc. London*, No. 8.

BRADY, G S, CROSSKEY, H W, and ROBERTSON, D. 1874. *A monograph on the Post-Tertiary entomostraca of Scotland.* (London: Palaeontographical Society.)

BROWNE, M A E, and GRAHAM, D K. 1981. Glaciomarine deposits of the Loch Lomond Stade glacier in the Vale of Leven between Dumbarton and Balloch, west-central Scotland. *Quaternary Newsletter* No. 34, 1–7.

— and MCMILLAN, A A. 1984. Shoreline inheritance and coastal history in the Firth of Clyde. *Scott. J. Geol.*, Vol. 20, 119–120.

— — 1985. The tills of Central Scotland in their stratigraphical context. In *Glacial tills 85*. FORDE, M C (editor). Proceedings of the International Conference on Construction in Glacial Tills and Boulder Clays. (Engineering Technics Press.)

— HARKNESS, D D, PEACOCK, J D, and WARD, R G. 1977. The date of deglaciation of the Paisley-Renfrew area. *Scott. J. Geol.*, Vol. 13, 301–303.

— MCMILLAN, A A, and GRAHAM, D K. 1983. A late-Devensian marine and non-marine sequence near Dumbarton, Strathclyde. *Scott. J. Geol.*, Vol. 19, 229–234.

— GRAHAM, D K, and GREGORY, D M. 1984. Quaternary estuarine deposits in the Grangemouth area, Scotland. *Rep. Inst. Geol. Sci.*, Vol. 16, No. 3.

— MCMILLAN, A A, and HALL, I H S. 1984. Blocks of marine clay in till near Helensburgh, Strathclyde. *Scott. J. Geol.*, Vol. 19, 321–325.

CHISHOLM, J I, and DEAN, J M. 1974. The Upper Old Red Sandstone of Fife and Kinross: a fluviatile sequence with evidence of marine incursion. *Scott. J. Geol.*, Vol. 10, 1–30.

COOMBS, D S, and WILKINSON, J F G. 1969. Lineages and fractionation trends in undersaturated volcanic rocks from the East Otago Volcanic Province (New Zealand) and related rocks. *J. Petrology*, Vol. 10, 440–501.

COOPE, G R. 1977. Fossil coleopteran assemblages as sensitive indicators of climatic changes during the Devensian (last) cold stage. *Philos. Trans. R. Soc. London*, Vol. B280, 313–340.

CURRY, G B, BLUCK, B J, BURTON, C J, INGHAM, J K, SIVETER, D J, and WILLIAMS, A. 1984. Age, evolution and tectonic history of the Highland Border Complex, Scotland. *Trans. R. Soc. Edinburgh: Earth Sci.*, Vol. 75, 113–133.

DEEGAN, C E, KIRBY, R, RAE, I, and FLOYD, R. 1973. The superficial deposits of the Firth of Clyde and its sea lochs. *Rep. Inst. Geol. Sci.*, No 73/9.

DE SOUZA, H A F. 1979. The geochronology of Scottish Carboniferous volcanism. Unpublished PhD thesis, University of Edinburgh.

ELLIOT, R W. 1985. Central Scotland Mineral Portfolio: Resources of clay and mudstone for brickmaking. *Open-File Report, Br. Geol. Surv., Edinburgh.*

EYLES, V A, SIMPSON, J B, and MACGREGOR, M C. 1949. Geology of Central Ayrshire. *Mem. Geol. Surv. G. B.*, Sheet 14 (Scotland).

FENTON, M W, and WILSON, C J L. 1985. Shallow-water turbidites: an example from the Mallacoota Beds, Australia. *Sediment. Geol.*, Vol. 45, 231–260.

FORD, W E. 1917. Studies in the calcite group. *Connecticut Acad. Arts & Sci.*, Vol. 22, 211–248.

FORSYTH, I H. 1978. The Lower Carboniferous sequence in the Howwood Syncline, Renfrewshire. *Bull. Geol. Surv. G. B.*, No. 60, 1–8.

— and WILSON, R B. 1965. Recent sections in the Lower Carboniferous of the Glasgow area. *Bull. Geol. Surv. G. B.*, No. 22, 65–79.

FRANCIS, E H, FORSYTH, I H, READ, W A, and ARMSTRONG, M. 1970. The geology of the Stirling district. *Mem. Geol. Surv. G. B.*

FRIEND, P F, HARLAND, W B, and HUDSON, J D. 1963. The Old Red Sandstone and the Highland Boundary in Arran, Scotland. *Trans. Edinburgh Geol. Soc.*, Vol. 19, 363–425.

FRYBERGER, S G, and SCHENK, C J. 1988. Pin-stripe lamination: a distinctive feature of modern and ancient eolian sediments. *Sediment. Geol.*, Vol. 55, 1–15.

GEORGE, T N. 1960. The stratigraphical evolution of the Midland Valley. *Trans. Geol. Soc. Glasgow*, Vol. 24, 32–107.

GRAHAM, D K, and WILKINSON, I P. 1978. A detailed examination of a late-Glacial faunal succession at Ardyne, Argyll, Scotland. *Rep. Inst. Geol. Sci.*, No. 78/5.

GRAY, J M. 1978. Low-level shore platforms in the south-west Scottish Highlands, age and correlation. *Trans. Inst. Br. Geogr.*, New Series 3, 151–164.

GUNN, W, CLOUGH, C T, and HILL, J B. 1897. The geology of Cowal. *Mem. Geol. Surv. G. B.*

HALL, I H S, and CHISHOLM, J I. 1987. Aeolian sediments in the Late Devonian of the Scottish Midland Valley. *Scott. J. Geol.*, Vol. 23, 203–208.

— and FORSYTH, I H. In preparation. Solid geology of the western and central parts of the Glasgow district. *Mem. Br. Geol. Surv.* Sheet 30E, (Scotland).

HARTE, B, BOOTH, J E, DEMPSTER, T J, FETTES, D J, MENDUM, J R, and WATTS, D. 1984. Aspects of the post-depositional evolution of Dalradian and Highland Border Complex rocks in the southern Highlands of Scotland. *Trans. R. Soc. Edinburgh: Earth Sci.*, Vol. 75, 151–163.

HERRIOT, A. 1971. A xenolithic dyke at M'Inroy's Point, Gourock, Renfrewshire. *Scott. J. Geol.*, Vol. 7, 153–161.

HINXMAN, L W, ANDERSON, E M, and CARRUTHERS, R G. 1920. The economic geology of the Central Coalfield of Scotland, area IV. *Mem. Geol. Surv. G. B.*

HOBSON, G V. 1959. Barytes in Scotland with special reference to Gasswater and Muirshiel mines. In *The future of non-ferrous mining in Great Britain and Ireland.* (London: Institute of Mining and Metallurgy.)

INSTITUTE OF GEOLOGICAL SCIENCES. 1977. IGS boreholes 1976. *Rep. Inst. Geol. Sci.*, No. 77/10.

— 1978. IGS boreholes 1977. *Rep. Inst. Geol. Sci.*, No. 78/21.

— 1979. IGS boreholes 1978. *Rep. Inst. Geol. Sci.*, No. 79/12.

— 1982. IGS boreholes 1980. *Rep. Inst. Geol. Sci.*, No. 81/11.

JARDINE, W G (editor). 1980. Glasgow region. *Quaternary Res. Assoc. Field Handbook.*

JOHNSTONE, G S. 1965. The volcanic rocks of the Misty Law-Knockside Hills district, Renfrewshire. *Bull. Geol. Surv. G. B.*, No. 22, 53–64.

KENNEDY, W Q. 1931. On composite lava flows. *Geol. Mag.*, Vol. 68, 166–181.

— 1933. Composite auto-intrusion in a Carboniferous lava flow. *Summary of progress for 1932.* GEOLOGICAL SURVEY OF GREAT BRITAIN. (London: Her Majesty's Stationery Office.)

— 1958. The tectonic evolution of the Midland Valley of Scotland. *Trans. Geol. Soc. Glasgow*, Vol. 23, 106–133.

KRZYWINSKI, K, and STABELL, B. 1984. Late-Weichselian sea level changes at Sotra, Hordaland, western Norway. *Boreas*, Vol. 13, 243–258.

LEITCH, P A, and SCOTT, A. 1917. Notes on the intrusive rocks of west Renfrewshire. *Trans. Geol. Soc. Glasgow*, Vol. 16, 275–289.

LOWE, J J, and GRAY, J M. 1979. The stratigraphic subdivision of the Lateglacial of NW Europe: a discussion. In *Studies in the Lateglacial of north-west Europe.* LOWE, J J, GRAY, J M, and ROBINSON, J E (editors). (Oxford and New York: Pergamon Press.)

MACDONALD, R. 1975. Petrochemistry of the early Carboniferous (Dinantian) lavas of Scotland. *Scott. J. Geol.*, Vol. 11, 269–314.

— GOTTFRIED, D, FARRINGTON, M J, BROWN, F W, and SKINNER, N G. 1981. The geochemistry of a continental tholeiitic suite: late Palaeozoic quartz-dolerite dykes of Scotland. *Trans. R. Soc. Edinburgh: Earth Sci.*, Vol. 72, 57–74.

MACGREGOR, A G. 1928. The classification of Scottish Carboniferous olivine-basalts and mugearites. *Trans. Geol. Soc. Glasgow*, Vol. 18, 324–360.

— 1944. Barytes in Central Scotland. *Wartime Pam. Geol. Surv. G. B.*, No. 38.

MERRITT, J W, and ELLIOT, R W. 1984. Central Scotland Mineral Portfolio: hard rock aggregate resources. *Open-file Report, Br. Geol. Surv.*, Edinburgh.

MIDDLETON, G V, and HAMPTON, M A. 1973. Sediment gravity flows; mechanics of flow and deposition. *In* Turbidites and deep-water sedimentation. *Soc. Econ. Paleontol. Mineral., Pac. Sect.*, 1–38.

MOORE, D J. 1979a. The baryte deposits of central and southern Scotland. Unpublished PhD thesis, University of London.

— 1979b. An unusual calcite from Muirshiel, Renfrewshire. *Mineral. Mag.*, Vol. 43, 446–448.

MORTON, D J. 1979. Palaeogeographical evolution of the Lower Old Red Sandstone basin in the western Midland Valley. *Scott. J. Geol.*, Vol. 15, 97–116.

MYKURA, W. 1983. Old Red Sandstone. In *Geology of Scotland* (2nd edition). CRAIG, G Y (editor). (Edinburgh: Scottish Academic Press.)

MUTTI, E, and RICCI-LUCCHI, F. 1972. Le torbiditi dell' Appenino settentrionale: introduzione all' analise di facies. *Mem. Soc. Geol. Ital.*, Vol. 11, 161–199.

NICHOLSON, K, and DURANT, G P. 1986. Fluorite-baryte-dolomite-iron-manganese mineralisation at Craigmuschat quarry; a historical review. *Scott. J. Geol.*, Vol. 22, 298–302.

PATERSON, I B. 1974. The supposed Perth readvance in the Perth District. *Scott. J. Geol.*, Vol. 10, 53–66.

— ARMSTRONG, M, and BROWNE, M A E. 1981. Quaternary estuarine deposits in the Tay–Earn area, Scotland. *Rep. Inst. Geol. Sci.*, No. 81/7.

— and HALL, I H S. 1986. Lithostratigraphy of the late Devonian and early Carboniferous rocks in the Midland Valley of Scotland. *Rep. Br. Geol Surv.*, Vol. 18, No. 3.

PEACH, A M. 1909. Boulder distribution from Lennoxtown, Scotland. *Geol. Mag.*, Vol. 46, 26–31.

PEACOCK, J D. 1971. Marine shell radiocarbon dates and the chronology of deglaciation in western Scotland. *Nature, London, Phys. Sci.*, Vol. 230, 43–45.

— 1975. Scottish late- and post-Glacial marine deposits. In *Quaternary studies in north-east Scotland.* GEMMELL, A M D (editor). (University of Aberdeen: Department of Geography.)

— 1981. Scottish late-Glacial marine deposits and their environmental significance. In *The Quaternary in Britain.* NEALE, J, and FLENLEY, J (editors). (Oxford and New York: Pergamon Press.)

— GRAHAM, D K, ROBINSON, J E, and WILKINSON, I P. 1977. Evolution and chronology of late-Glacial environments at Lochgilphead, Scotland. In *Studies in the Scottish lateglacial environments.* GRAY, J M, and LOWE, J J (editors). (Oxford and New York: Pergamon Press.)

— — and WILKINSON, I P. 1978. Late-Glacial marine environments at Ardyne, Scotland, and their significance in the interpretation of the history of the Clyde sea area. *Rep. Inst. Geol. Sci.,* No. 78/17.

PENNINGTON, W. 1977. The late Devensian flora and vegetation of Britain. *Philos. Trans. R. Soc. London,* Vol. B280, 247–271.

PRICE, R J. 1975. The glaciation of west-central Scotland—a review. *Scott. Geogr. Mag.,* Vol. 91, 134–145.

— 1983. *Scotland's Environment during the Last 30 000 Years.* (Edinburgh: Scottish Academic Press.)

QURESHI, I R. 1970. A gravity survey in the region of the Highland Boundary Fault in Scotland. *Q. J. Geol. Soc. London,* Vol. 125, 481–502.

RAMSAY, D M. 1964. Deformation of pebbles in Lower Old Red Sandstone conglomerates adjacent to the Highland Boundary Fault. *Geol. Mag.,* Vol. 101, 228–248.

READ, W A, and JOHNSON, S R H. 1967. The sedimentology of sandstone formations within the Upper Old Red Sandstone and lowest Calciferous Sandstone Measures west of Stirling, Scotland. *Scott. J. Geol.,* Vol. 3, 242–267.

RICHEY, J E. 1928. The north Ayrshire sequence of Calciferous Sandstone volcanic rocks. *Trans. Geol. Soc. Glasgow.* Vol. 18, 247–255.

— 1939. The dykes of Scotland. *Trans. Edinburgh Geol. Soc.,* Vol. 13, 393–435.

— ANDERSON, E M, and MACGREGOR, A G. 1930. The geology of north Ayrshire. *Mem. Geol. Surv. G. B.*

ROBERTS, J L, and TREAGUS, J E. 1977. The Dalradian rocks of the south-west Highlands—introduction. *Scott. J. Geol.,* Vol. 13, 87–99.

ROBERTSON, D. 1883. On the post-Tertiary beds of Garvel Park, Greenock. *Trans. Geol. Soc. Glasgow,* Vol. 7, 297–309.

ROGERS, G, DEMPSTER, T J, BLUCK, B J, and TANNER, P W G. 1989. A high precision U-Pb age for the Ben Vuirich granite: implications for the evolution of the Scottish Dalradian Supergroup. *J Geol. Soc. London,* Vol. 146, 789–798.

ROSE, J. 1975. Raised beach gravels and ice-wedge casts at Old Kilpatrick, near Glasgow. *Scott. J. Geol.,* Vol. 11, 15–21.

RUDDIMAN, W F, and MCINTYRE, A. 1973. Time-transgressive deglacial retreat of polar waters from the North Atlantic. *Quaternary Res.,* Vol. 3, 128–140.

RUSSELL, M J, and SMYTH, D K. 1983. Origin of the Oslo Graben in relation to the Hercynian-Alleghenian Orogeny and lithospheric rifting in the North Atlantic. *In* Developments in geotectonics, 19: Processes of continental rifting. MORGAN, P, and BAKER, B E (editors). *Tectonophysics,* Vol. 94, 457–472.

SCOTT, A C, EDWARDS, D, and ROLFE, W D I. 1976. Fossiliferous Lower Old Red Sandstone near Cardross, Dunbartonshire. *Proc. Geol. Soc. Glasgow,* Vol. 117, 4–5.

SCOTT, T, and STEEL, J. 1883. Notes on the occurrence of *Leda arctica* (Gray); *Lyonsia arenosa* (Möller), and other organic remains, in the post-Pliocene clays of Garvel Park, Greenock. *Trans. Geol. Soc. Glasgow,* Vol. 7, 274–283.

SIMPSON, J C. 1933. The late-Glacial readvance moraines of the Highland Border west of the River Tay. *Trans. R. Soc. Edinburgh,* Vol. 57, 633–646.

SISSONS, J B. 1966. Relative sea-level changes between 10 300 and 8300 BP in part of the Carse of Stirling. *Trans. Inst. Br. Geogr.,* Vol. 39, 19–29.

— 1967. *The evolution of Scotland's Scenery.* (Edinburgh: Oliver and Boyd.)

— 1969. Drift stratigraphy and buried morphological features in the Grangemouth–Falkirk–Airth area, central Scotland. *Trans. Inst. Br. Geogr.,* Vol. 48, 19–50.

— 1974. Lateglacial marine erosion in Scotland. *Boreas,* Vol. 3, 41–48.

— 1981. The late Scottish ice-sheet: facts and speculative discussion. *Boreas,* Vol. 10, 1–17.

— and BROOKS, C L. 1971. Dating of early postglacial land and sea-level changes in the western Forth Valley. *Nature, London,* Vol. 234, 126–127.

— and SMITH, D E. 1965. Raised shorelines associated with the Perth Readvance in the Forth valley and their relation to glacial isostasy. *Trans. R. Soc. Edinburgh,* Vol. 66, 143–168.

SMEDLEY, P L. 1986. Petrochemistry of Dinantian volcanism in northern Britain. Unpublished PhD thesis, University of Edinburgh.

STEPHENSON, D, and COATS, J S. 1983. Baryte and copper mineralisation in the Renfrewshire Hills, central Scotland. *Mineral Reconaissance Rep. Inst. Geol. Sci.,* No. 67.

STEWART, A S, WALKER, A, and DICKSON, J H. 1984. Pollen diagrams from Dubh Lochan, near Loch Lomond. *New Phytol.,* Vol. 98, 531–549.

SUTHERLAND, D G. 1979. Problems of radiocarbon dating deposits from newly deglaciated terrain: examples from the Scottish Lateglacial. In *Studies in the Lateglacial of North-west Europe.* LOWE, J J, GRAY, J M, and ROBINSON, J E (editors). (Oxford and New York: Pergamon Press.)

— 1984. The Quaternary deposits and landforms of Scotland and the neighbouring shelves: a review. *Quaternary Sci. Rev.,* Vol. 3, 157–254.

THOMPSON, R N. 1982. Geochemistry and magma genesis. In *Igneous rocks of the British Isles.* SUTHERLAND, D S (editor). (Chichester: Wiley.)

TYRRELL, G W. 1917. The trachytic and allied rocks of the Clyde Carboniferous lava plateau. *Proc. R. Soc. Edinburgh,* Vol. 36, 288–299.

WALKER, F. 1935. The late Palaeozoic quartz-dolerites and tholeiites of Scotland. *Mineral. Mag.,* Vol. 24, 131–159.

WHYTE, F. 1966. Dumbarton Rock. *Scott. J. Geol.,* Vol. 2, 107–121.

— 1980. Trace element variations at the contact of the Dumbarton Rock basalt. *Scott. J. Geol.,* Vol. 16, 263–266.

— and MACDONALD, J G. 1974. Lower Carboniferous vulcanicity in the northern part of the Clyde Plateau. *Scott. J. Geol.,* Vol. 10, 187–198.

WILSON, G V. 1921. The Pb, Zn, Cu and Ni ores of Scotland. *Spec. Rep. Miner. Resour. Mem. Geol. Surv. G. B.,* No. 17.

56 REFERENCES

— EASTWOOD, T, POCOCK, R W, WRAY, D A, and ROBERTSON,
T. 1922. Barytes and Witherite. *Spec. Rep. Miner. Resour.
Mem. Geol. Surv. G. B.*, No. 22.

WILSON, R B. 1966. A study of the Neilson Shell Bed, a
Scottish Lower Carboniferous marine shale. *Bull. Geol. Surv.
G. B.*, No. 24, 105 – 130.

APPENDIX 1

Quaternary fossils from the Greenock district

The Table includes only taxa identified in available material from the district and does not purport to be a fully comprehensive faunal list for the formations shown. No Flandrian microfaunas from this area have been examined. (x taxon recorded • taxon not recorded)

MICROFAUNA

FORAMINIFERIDA

	MLD	PF	LF	BF
Ammonia batavus	.	.	.	x
Angulogerina fluens	.	.	.	x
Bolivina sp.	.	.	x	x
Bucella frigida	.	.	x	x
B. marginata	.	.	.	x
Cassidulina reniforme	x	x	.	x
Cibicides lobatulus	x	.	x	x
Elphidium albiumbilicatum	x	.	x	x
E. asklundi	.	.	.	x
E. bartletti	.	.	.	x
E. clavatum	x	x	x	x
E. subarcticum	.	.	.	x
E. williamsoni	.	.	x	.
Fissurina laevigata	.	.	x	.
F. lucida	.	.	x	.
F. marginata	.	.	.	x
F. orbignyana	.	.	x	.
Guttulina?	.	.	x	.
Haynesina germanica	.	.	x	.
H. orbiculare	.	.	x	.
Lagena clavata	.	.	x	.
L. distoma	.	.	x	x
L. laevis	.	.	x	x
L. striata	.	.	x	.
L. substriata	.	.	x	.
Miliammina fusca	.	.	x	.
Miliolinella subrotunda	.	.	x	x
M. sp.	.	.	x	x
Nonion anglicum	.	.	.	x
N. orbiculare	x	.	.	x
Patellina corrugata	x	.	.	x
Pyrgo williamsoni	x	x	x	x
Quinqueloculina seminulum	.	x	x	x
Q. sp.	.	.	.	x
Virgulina loeblichi	.	.	x	.
polymorphinids	.	.	x	x

OSTRACODA

	MLD	PF	LF	BF
Acanthocythereis dunelmensis	x	.	x	x
Bythocythere constricta	.	.	x	.
Cluthia cluthae	.	.	x	.
Cyprideis torosa	.	.	x	.
Cytherois fischeri	.	.	x	.
Cytheropteron biconvexa	.	.	.	x
C. latissimum	.	.	x	x
C. nodosum	.	.	x	.
Eucytheridea macrolaminata	.	.	.	x
Hirschmannia tamarindus	.	.	x	.
H. viridis	.	.	x	.
Jonesia simplex	.	.	x	.
Leptocythere castanea	.	.	x	.
Paradoxostoma variabile	.	.	x	.
Polycope orbiculare	.	.	x	.
Pontocypris mytilloides	.	.	x	.
Robertsonites tuberculata	.	.	x	.
Sarsicytheridea punctillata	.	.	x	.
Sclerochilus contortus	.	.	x	.
Semicytherura nigrescens	.	.	x	.
Xestoleberis depressa	.	.	x	.

MACROFAUNA

	MLD	PF	LF	BF	F
ANNELIDA					
Hydroides norvegica	.	.	x	.	
GASTROPODA					
Acmaea virginea	.	.	x	.	
Boreotrophon truncatus	.	.	x	.	
Cylichnina cylindracea	.	.	x	.	
Gibbula cineraria	.	.	.	.	x
Lacuna vincta	.	.	x	.	x
Littorina littorea	.	.	x	.	x
L. obtusata	.	.	x	.	
L. rudis	.	.	x	.	
L. spp.	.	.	.	.	
Lunatia alderi	.	.	.	.	x
L. pallida	.	.	x	.	x
Nassarius incrassatus	.	.	.	.	.
Nucella lapillus	.	.	x	.	x
Oenopota cf. *angulosa*	.	.	x	.	

	MLD	PF	LF	BF	F
O. turricula	.	.	x	.	.
Onoba semicostata	.	.	x	.	.
Philbertia linearis	.	.	.	.	x
Puncturella noachina	.	.	x	.	x
Retusa obtusa	.	.	x	.	.
R. truncatula	.	.	.	.	.
R. umbilicata	.	.	.	.	x
Rissoa parva interrupta	.	.	.	.	x
R. cf. rufilabrum	.	.	.	.	.
R. sarsi	.	.	.	.	x
Skeneopsis planorbis	.	.	x	.	x
Turbonilla elegantissima	.	.	.	.	.
Turritella communis	.	.	.	.	x
BIVALVIA					
Abra alba	.	.	.	.	.
Acanthocardia echinata	.	.	x	.	x
Arca pectunculoides	.	.	.	.	x
Arctica islandica	.	.	x	.	.
Cerastoderma edule	.	.	x	.	.
Chlamys opercularis	.	.	.	.	x
C. islandica	x	x	.	.	.
Cochlodesma praetenue	.	.	.	.	x
Corbula gibba	.	.	.	.	x
Dosinia exoleta	.	.	.	.	x
Fabulina fabula	.	.	.	.	x
Heteranomia squamula	.	.	x	.	.
Hiatella arctica	.	.	.	x	x
Laevicardium sp.	.	.	.	.	x
Lucinoma borealis	.	.	.	.	x
Macoma balthica?	.	.	x	.	.
M. calcarea	.	.	x	.	.
Modiolus modiolus	.	.	x	.	.
Montacuta ferruginosa	.	.	.	.	x
Musculus discors	.	.	x	.	.
M. sp.	.	.	.	x	.
Mya arenaria	.	.	x	.	.

	MLD	PF	LF	BF	F
M. truncata	.	.	x	.	.
Mysella bidentata	.	.	.	.	x
Mytilus edulis	.	.	x	.	x
Nicania montagui	.	.	x	.	x
Nucula tenuis	.	.	x	.	.
N. sp.	.	.	.	.	x
Nuculana minuta	.	.	x	.	.
N. pernula	.	.	x	.	.
Parvicardium ovale	.	.	x	x	x
P. sp.	.	.	.	.	x
Portlandia arctica	x	.	.	x	.
Spisula elliptica	.	.	.	.	x
S. subtruncata	.	.	.	.	x
Thracia?	.	.	.	.	x
Thyasira cf. gouldi	.	x	.	.	.
T. flexuosa	.	.	.	.	x
T. sp.	x	.	x	.	.
Tridonta elliptica	.	.	x	.	x
T. sp.	.	.	x	.	x
Venus fasciata	.	.	.	.	x
V. ovata	.	.	.	.	x
V. striatula	.	.	.	.	x
Yoldiella fraterna	.	.	x	.	.
Y. lenticula	.	.	x	x	.
mytilacean fragments	x	.	.	x	.
solenacean fragments	.	.	.	.	x
BALANOMORPHIDA					
Balanus sp. plates	.	x	.	.	x
Verruca sp. plates	.	x	.	.	.
ECHINODERMATA					
Echinus sp. plates	.	.	.	.	x
echinoid spines	.	x	x	x	.
starfish fragments	.	.	x	x	.

Note
MLD   Main late Devonian
PF   Paisley Formation
LF   Linwood Formation
BF   Balloch Formation
F   Flandrian

APPENDIX 2

List of BGS boreholes in solid rocks of the Greenock district and adjacent areas

Name	Accession number in BGS files	Grid reference
Barnhill	NS 47 NW/2	[4269 7571]
Everton	NS 27 SW/5	[2145 7103]
Glenburn	NS 46 SE/164	[4783 6065]
Kipperoch	NS 37 NE/20	[3727 7742]
Knocknairshill	NS 37/SW/10	[3056 7438]
Largs	NS 25 NW/5	[2158 5936]
Loch Humphrey	NS 47 NE/1	[4582 7555]
Lora Burn	NS 35 NW/51	[3336 5841]

Preliminary logs of these boreholes have been published in IGS Report Nos. 77/10; 78/21; 79/12; 81/11.

APPENDIX 3

Summary logs of BGS drift boreholes located in or near the Greenock district

Depths and thicknesses are in metres

Balloch Borehole [38905 81955] Surface level 11.05 m above OD

	Thickness m	Depth m
QUATERNARY		
Made ground	1.00	1.00
Gravel, fine to coarse, with shell fragments	4.50	5.50
Sand, fine to coarse, some fine gravel	5.00	10.50
Clay, very stiff to hard, grey and brownish grey, with bands of fine to medium sand and clayey silt; shell fragments at 24.00 m and below 45.75 m	39.75	50.25
Gravel, shelly, fine to coarse	0.45	50.70
Till, stiff reddish brown stony clay	1.80	52.50
LOWER DEVONIAN		
Sandstone, medium- to fine-grained, dark grey and greenish grey	0.15	52.65
		End of borehole

Erskine Bridge Borehole [46345 72510] Surface level 3.19 m above OD

	Thickness m	Depth m
QUATERNARY		
Made ground	4.25	4.25
Silt, grey, clayey	3.25	7.50
Gravel, shelly	2.52	10.02
Till (upper), reddish brown	5.98	16.00
Clay, laminated, reddish brown, over-consolidated	16.75	32.75

	Thickness m	Depth m
Till (lower), reddish brown	1.25	34.00
		End of borehole

Inverleven (Dumbarton) Borehole [39752 75089] Surface level 3.98 m above OD

	Thickness m	Depth m
QUATERNARY		
Made ground	8.50	8.50
Clay, brownish grey and grey with bands of silty clay, silt and fine sand; scattered shells	45.50	54.00
Gravel, shelly, fine to very coarse	0.80	54.80
Silt, greyish green, with bands of fine to medium sand; rare shell fragments	11.20	66.00
LOWER CARBONIFEROUS		
Sandstone, fine- to medium-grained, red and white mottled	1.10	67.10
		End of borehole

Linwood Borehole [4459 6588] Surface level 10.18 m above OD

	Thickness m	Depth m
QUATERNARY		
Peat	3.50	3.50
Silt, clayey, with bands of silty clay, grey and brown, colour-laminated towards base	27.60	31.10
Till, stiff stony clay	0.90	32.00
		End of borehole

The boreholes were drilled in 1980 and logged by M A E Browne, D N Halley and K I G Lawrie (Institute of Geological Sciences, 1982). The results of the boreholes were discussed by Browne (*in* Jardine, 1980) and more detailed accounts of the Balloch and Inverleven bores were given by Browne and Graham (1981).

APPENDIX 4

Six-inch maps (Solid)

The maps at six inches to one mile covering, wholly or in part, the solid rocks in 1:50 000 sheet 30W and the eastern part of sheet 29E are listed below with the names of the surveyors (M Armstrong, I B Cameron, I H Forsyth, I H S Hall, W G Henderson, S K Monro, I B Paterson and D Stephenson) and the date of survey. The maps are not published but most of them are available for consultation in the British Geological Survey office, Murchison House, Edinburgh where photocopies can be purchased.

NATIONAL GRID SHEETS

NS 16 NE	Paterson	1977–78
NS 16 SE	Cameron and Armstrong	1981
NS 17 SE	Paterson	1977
NS 25 NW	Stephenson	1982
NS 25 NE	Stephenson	1979–80
NS 26 NW	Paterson and Stephenson	1976–82
NS 26 NE	Stephenson	1979–81
NS 26 SW	Stephenson and Cameron	1981–82
NS 26 SE	Stephenson	1979–80
NS 27 NW	Paterson	1976
NS 27 NE	Paterson	1976–77
NS 27 SW	Paterson	1976–78
NS 27 SE	Paterson	1976–81
NS 28 NW	Henderson	1977–78
NS 28 NE	Henderson	1979
NS 28 SW	Henderson	1981
NS 28 SE	Henderson and Hall	1977–81
NS 35 NW	Stephenson and Monro	1978–81
NS 35 NE	Stephenson and Monro	1978–80
NS 36 NW	Paterson and Stephenson	1977–81
NS 36 NE	Paterson	1978–79
NS 36 SW	Stephenson	1979–81
NS 36 SE	Stephenson	1980
NS 37 NW	Hall and Paterson	1977–81
NS 37 NE	Hall	1977–82
NS 37 SW	Paterson	1977
NS 37 SE	Paterson and Hall	1977–79
NS 38 NW	Henderson and Hall	1977–81
NS 38 NE	Hall	1981
NS 38 SW	Hall	1980–82
NS 38 SE	Hall	1977–82
NS 45 NW	Stephenson	1979
NS 46 NW	Paterson and Forsyth	1979–80
NS 46 SW	Paterson, Stephenson and Forsyth	1979–80
NS 47 NW	Hall	1974–78
NS 47 SW	Paterson and Hall	1977–80
NS 48 NW	Hall	1980
NS 48 SW	Hall	1977–79

APPENDIX 5

Six-inch maps (Drift)

The field maps at six inch to one mile covering, wholly or in part the superficial deposits in 1:50 000 sheet 30W and the eastern part of sheet 29E are listed below with the names of the surveyors (J G C Anderson, I B Cameron, I H S Hall, W G Henderson, S K Monro, I B Paterson, J E Richey, J B Simpson and D Stephenson) and the date of survey.

The maps are not published but most of them are available in the British Geological Survey office, Murchison House, Edinburgh where photocopies can be purchased.

NATIONAL GRID SHEETS

NS 16 NE	Paterson	1977 – 78
NS 16 SE	Cameron	1981
NS 17 SE	Paterson	1977
NS 25 NW	Stephenson	1982
NS 25 NE	Stephenson	1979 – 80
NS 26 NW	Paterson and Stephenson	1976 – 82
NS 26 NE	Stephenson	1979 – 81
NS 26 SW	Stephenson and Cameron	1981 – 82
NS 26 SE	Stephenson	1979 – 80
NS 27 NW	Paterson	1976
NS 27 NE	Paterson	1976 – 77
NS 27 SW	Paterson	1976 – 78
NS 27 SE	Paterson	1976 – 81
NS 28 NW	Cameron	1977 – 78
NS 28 SW	Henderson and Cameron	1981
NS 35 NW	Stephenson and Monro	1978 – 81
NS 35 NE	Stephenson and Monro	1978 – 80
NS 36 NW	Paterson and Stephenson	1977 – 81
NS 36 NE	Paterson	1978 – 79
NS 36 SW	Stephenson	1979 – 81
NS 36 SE	Stephenson	1980
NS 37 NW	Paterson	1977
NS 37 SW	Paterson	1977
NS 37 SE	Paterson and Hall	1977 – 79
NS 45 NW	Stephenson	1979
NS 46 NW	Paterson	1979 – 80
NS 46 SW	Paterson and Stephenson	1979 – 80
NS 47 SW	Paterson and Hall	1977 – 80

COUNTY SHEETS (NEW MERIDAN)

Dunbarton 12 NE	Anderson	1947
Dunbarton 12 SE	Anderson	1946
Dunbarton 13 NW	Anderson	1947
Dunbarton 13 NE	Anderson and Richey	1937 – 47
Dunbarton 13 SW	Anderson and Richey	1937 – 47
Dunbarton 13 SE	Anderson and Richey	1937 – 47
Dunbarton 14 NW	Anderson and Richey	1937 – 47
Dunbarton 14 NE	Anderson (1)	1939
Dunbarton 14 SW	Anderson and Richey (1)	1937 – 47
Dunbarton 14 SE	Anderson and Simpson	1927 – 39
Dunbarton 16 NE	Anderson	1946
Dunbarton 17 NW	Anderson and Richey	1939 – 46
Dunbarton 17 NE	Richey	1939
Dunbarton 17 SE	Anderson, Richey and Simpson	1929 – 47
Dunbarton 18 NW	Anderson, Richey and Simpson (2; 3)	1928 – 47
Dunbarton 18 NE	Anderson and Simpson (2; 3)	1929 – 39
Dunbarton 18 SW	Anderson, Richey and Simpson (3)	1928 – 47
Dunbarton 18 SE	Anderson (3)	1938
Dunbarton 22 NW	Anderson (2)	1938 – 47
Dunbarton 22 NE	Anderson	1938

Amendments to the above County maps by A M Aitken (1), M A E Browne (2) and I H S Hall (3) have been included in the 1:50 000 maps.

INDEX

BRITISH GEOLOGICAL SURVEY

Keyworth, Nottingham NG12 5GG
(06077) 6111

Murchison House, West Mains Road,
Edinburgh EH9 3LA   031-667 1000

London Information Office, Natural History Museum
Earth Galleries, Exhibition Road, London SW7 2DE
071-589 4090

The full range of Survey publications is available
through the Sales Desks at Keyworth, Murchison
House, Edinburgh, and at the BGS London Information Office in the Natural History Museum, Earth
Galleries. The adjacent bookshop stocks the more
popular books for sale over the counter. Most BGS
books and reports are listed in HMSO's Sectional List
45, and can be bought from HMSO and through
HMSO agents and retailers. Maps are listed in the
BGS Map Catalogue and the Ordnance Survey's Trade
Catalogue, and can be bought from Ordnance Survey
agents as well as from BGS.

The British Geological Survey carries out the geological survey of
Great Britain and Northern Ireland (the latter as an agency
service for the government of Northern Ireland), and of the
surrounding continental shelf, as well as its basic research
projects. It also undertakes programmes of British technical aid
in geology in developing countries as arranged by the Overseas
Development Administration.

The British Geological Survey is a component body of the
Natural Environment Research Council.

Maps and diagrams in this book use topography based
on Ordnance Survey mapping

HMSO publications are available from:

HMSO Publications Centre
(Mail and telephone orders)
PO Box 276, London SW8 5DT
Telephone orders 071-873 9090
General enquiries 071-873 0011
Queueing system in operation for both numbers

HMSO Bookshops
49 High Holborn, London WC1V 6HB
    071-873 0011 (Counter service only)
258 Broad Street, Birmingham B1 2HE
    021-643 3740
Southey House, 33 Wine Street, Bristol BS1 2BQ
    (0272) 264306
9 Princess Street, Manchester M60 8AS
    061-834 7201
80 Chichester Street, Belfast BT1 4JY
    (0232) 238451
71 Lothian Road, Edinburgh EH3 9AZ
    031-228 4181

HMSO's Accredited Agents
(see Yellow Pages)

And through good booksellers